MW00990122

Graduate Texts in Mathematics 177

Springer
New York
Berlin
Heidelberg
Barcelona
Budapest
Hong Kong
London
Milan
Paris
Santa Clara
Singapore
Tokyo

Graduate Texts in Mathematics

1 TAKEUTI/ZARING. Introduction to Axiomatic Set Theory. 2nd ed.
2 OXTOBY. Measure and Category. 2nd ed.
3 SCHAEFER. Topological Vector Spaces.
4 HILTON/STAMMBACH. A Course in Homological Algebra. 2nd ed.
5 MAC LANE. Categories for the Working Mathematician.
6 HUGHES/PIPER. Projective Planes.
7 SERRE. A Course in Arithmetic.
8 TAKEUTI/ZARING. Axiomatic Set Theory.
9 HUMPHREYS. Introduction to Lie Algebras and Representation Theory.
10 COHEN. A Course in Simple Homotopy Theory.
11 CONWAY. Functions of One Complex Variable I. 2nd ed.
12 BEALS. Advanced Mathematical Analysis.
13 ANDERSON/FULLER. Rings and Categories of Modules. 2nd ed.
14 GOLUBITSKY/GUILLEMIN. Stable Mappings and Their Singularities.
15 BERBERIAN. Lectures in Functional Analysis and Operator Theory.
16 WINTER. The Structure of Fields.
17 ROSENBLATT. Random Processes. 2nd ed.
18 HALMOS. Measure Theory.
19 HALMOS. A Hilbert Space Problem Book. 2nd ed.
20 HUSEMOLLER. Fibre Bundles. 3rd ed.
21 HUMPHREYS. Linear Algebraic Groups.
22 BARNES/MACK. An Algebraic Introduction to Mathematical Logic.
23 GREUB. Linear Algebra. 4th ed.
24 HOLMES. Geometric Functional Analysis and Its Applications.
25 HEWITT/STROMBERG. Real and Abstract Analysis.
26 MANES. Algebraic Theories.
27 KELLEY. General Topology.
28 ZARISKI/SAMUEL. Commutative Algebra. Vol.I.
29 ZARISKI/SAMUEL. Commutative Algebra. Vol.II.
30 JACOBSON. Lectures in Abstract Algebra I. Basic Concepts.
31 JACOBSON. Lectures in Abstract Algebra II. Linear Algebra.
32 JACOBSON. Lectures in Abstract Algebra III. Theory of Fields and Galois Theory.
33 HIRSCH. Differential Topology.
34 SPITZER. Principles of Random Walk. 2nd ed.
35 WERMER. Banach Algebras and Several Complex Variables. 2nd ed.
36 KELLEY/NAMIOKA et al. Linear Topological Spaces.
37 MONK. Mathematical Logic.
38 GRAUERT/FRITZSCHE. Several Complex Variables.
39 ARVESON. An Invitation to C^*-Algebras.
40 KEMENY/SNELL/KNAPP. Denumerable Markov Chains. 2nd ed.
41 APOSTOL. Modular Functions and Dirichlet Series in Number Theory. 2nd ed.
42 SERRE. Linear Representations of Finite Groups.
43 GILLMAN/JERISON. Rings of Continuous Functions.
44 KENDIG. Elementary Algebraic Geometry.
45 LOÈVE. Probability Theory I. 4th ed.
46 LOÈVE. Probability Theory II. 4th ed.
47 MOISE. Geometric Topology in Dimensions 2 and 3.
48 SACHS/WU. General Relativity for Mathematicians.
49 GRUENBERG/WEIR. Linear Geometry. 2nd ed.
50 EDWARDS. Fermat's Last Theorem.
51 KLINGENBERG. A Course in Differential Geometry.
52 HARTSHORNE. Algebraic Geometry.
53 MANIN. A Course in Mathematical Logic.
54 GRAVER/WATKINS. Combinatorics with Emphasis on the Theory of Graphs.
55 BROWN/PEARCY. Introduction to Operator Theory I: Elements of Functional Analysis.
56 MASSEY. Algebraic Topology: An Introduction.
57 CROWELL/FOX. Introduction to Knot Theory.
58 KOBLITZ. p-adic Numbers, p-adic Analysis, and Zeta-Functions. 2nd ed.
59 LANG. Cyclotomic Fields.
60 ARNOLD. Mathematical Methods in Classical Mechanics. 2nd ed.

continued after index

Donald J. Newman

Analytic Number Theory

Donald J. Newman
National Security Agency
Fort Meade, MD 20755
USA

Mathematics Subject Classification (1991): 15A60, 65F35, 15A51

Library of Congress Cataloging-in-Publication Data
Newman, Donald J., 1930–
Analytic number theory / Donald J. Newman.
p. cm. — (Graduate texts in mathematics ; 177)
Includes index.
ISBN 0-387-98308-2 (hardcover : alk. paper)
1. Number Theory. I. Title. II. Series.
QA241.N48 1997
512.′73—dc21 97-26431

Printed on acid-free paper.

Production managed by Allan Abrams; manufacturing supervised by Jeffrey Taub.
Photocomposed pages prepared from the author's LaTeX files.
Printed and bound by Edwards Brothers, Inc., Ann Arbor, MI.
Printed in the United States of America.

9 8 7 6 5 4 3 2 1

ISBN 0-387-98308-2 Springer-Verlag New York Berlin Heidelberg SPIN 10635124

Contents

Introduction and Dedication vii

I. The Idea of Analytic Number Theory **1**

Addition Problems 1
Change Making 2
Crazy Dice 5
Can $r(n)$ be "constant?" 8
A Splitting Problem 8
An Identity of Euler's 11
Marks on a Ruler 12
Dissection into Arithmetic Progressions 14

II. The Partition Function **17**

The Generating Function 18
The Approximation 19
Riemann Sums 20
The Coefficients of $q(n)$ 25

III. The Erdös–Fuchs Theorem **31**

Erdös–Fuchs Theorem 35

IV. Sequences without Arithmetic Progressions **41**

The Basic Approximation Lemma 42

V. The Waring Problem **49**

VI. A "Natural" Proof of the Nonvanishing of *L*-Series **59**

VII. Simple Analytic Proof of the Prime Number Theorem **65**

First Proof of the Prime Number Theorem. 68
Second Proof of the Prime Number Theorem. 70

Index **75**

Introduction and Dedication

This book is dedicated to Paul Erdös, the greatest mathematician I have ever known, whom it has been my rare privilege to consider colleague, collaborator, and dear friend.

I like to think that Erdös, whose mathematics embodied the principles which have impressed themselves upon me as defining the true character of mathematics, would have appreciated this little book and heartily endorsed its philosophy. This book proffers the thesis that mathematics is actually an easy subject and many of the famous problems, even those in number theory itself, which have famously difficult solutions, can be resolved in simple and more direct terms.

There is no doubt a certain presumptuousness in this claim. The great mathematicians of yesteryear, those working in number theory and related fields, did not necessarily strive to effect the simple solution. They may have felt that the status and importance of mathematics as an intellectual discipline entailed, perhaps indeed required, a weighty solution. Gauss was certainly a wordy master and Euler another. They belonged to a tradition that undoubtedly revered mathematics, but as a discipline at some considerable remove from the commonplace. In keeping with a more democratic concept of intelligence itself, contemporary mathematics diverges from this somewhat elitist view. The simple approach implies a mathematics generally available even to those who have not been favored with the natural endowments, nor the careful cultivation of an Euler or Gauss.

Such an attitude might prove an effective antidote to a generally declining interest in pure mathematics. But it is not so much as incentive that we proffer what might best be called "the fun and games" approach to mathematics, but as a revelation of its true nature. The insistence on simplicity asserts a mathematics that is both "magical" and coherent. The solution that strives to master these qualities restores to mathematics that element of adventure that has always supplied its peculiar excitement. That adventure is intrinsic to even the most elementary description of analytic number theory.

The initial step in the investigation of a number theoretic item is the formulation of "the generating function". This formulation inevitably moves us away from the designated subject to a consideration of complex variables. Having wandered away from our subject, it becomes necessary to effect a return. Toward this end "The Cauchy Integral" proves to be an indispensable tool. Yet it leads us, inevitably, further afield to all the intricacies of contour integration and they, in turn entail the familiar processes, the deformation and estimation of these contour integrals.

Retracing our steps we find that we have gone from number theory to function theory, and back again. The journey seems circuitous, yet in its wake a pattern is revealed that implies a mathematics deeply inter-connected and cohesive.

I

The Idea of Analytic Number Theory

The most intriguing thing about Analytic Number Theory (the use of *Analysis*, or *function theory*, in number theory) is its very existence! How could one use properties of continuous valued functions to determine properties of those most discrete items, the integers. Analytic functions? What has differentiability got to do with counting? The astonishment mounts further when we learn that the complex zeros of a certain analytic function are *the* basic tools in the investigation of the primes.

The answer to all this bewilderment is given by the two words *generating functions*. Well, there are answers and answers. To those of us who have witnessed the use of generating functions this is a kind of answer, but to those of us who haven't, this is simply a restatement of the question. Perhaps the best way to understand the use of the analytic method, or the use of generating functions, is to see it in action in a number of pertinent examples. So let us take a look at some of these.

Addition Problems

Questions about addition lend themselves very naturally to the use of generating functions. The link is the simple observation that adding m and n is isomorphic to multiplying z^m and z^n. Thereby questions about the addition of integers are transformed into questions about the multiplication of polynomials or power series. For example, Lagrange's beautiful theorem that every positive integer is the sum of four squares becomes the statement that all of the coefficients of the power series for $\left(1 + z + z^4 + \cdots + z^{n^2} + \cdots\right)^4$ are positive. How one proves such a fact about the coefficients of such a power series is another story, but at least one begins to see how this transition from integers to analytic functions takes place. But now let's look at some addition problems that we *can* solve completely by the analytic method.

Change Making

How many ways can one make change of a dollar? The answer is 293, but the problem is both too hard and too easy. Too hard because the available coins are so many and so diverse. Too easy because it concerns just one "change," a dollar. More fitting to our spirit is the following problem: How many ways can we make change for n if the coins are 1, 2, and 3? To form the appropriate generating function, let us write, for $|z| < 1$,

$$\frac{1}{1-z} = 1 + z + z^{1+1} + z^{1+1+1} + \cdots,$$

$$\frac{1}{1-z^2} = 1 + z^2 + z^{2+2} + z^{2+2+2} + \cdots,$$

$$\frac{1}{1-z^3} = 1 + z^3 + z^{3+3} + z^{3+3+3} + \cdots,$$

and multiplying these three equations to get

$$\frac{1}{(1-z)(1-z^2)(1-z^3)} = (1 + z + z^{1+1} + \cdots)(1 + z^2 + z^{2+2} + \cdots) \times (1 + z^3 + z^{3+3} + \cdots).$$

Now we ask ourselves What happens when we multiply out the right-hand side? We obtain terms like $z^{1+1+1+1} \cdot z^2 \cdot z^{3+3}$. On the one hand, this term is z^{12}, but, on the other hand, it is $z^{\text{four 1's} + \text{one 2} + \text{two 3's}}$ and doesn't this exactly correspond to the method of changing the amount 12 into four 1's, one 2, and two 3's? Yes, and in fact we see that "every" way of making change (into 1's, 2's, and 3's) for "every" n will appear in this multiplying out. Thus if we call $C(n)$ the number of ways of making change for n, then $C(n)$ will be the exact coefficient of z^n when the multiplication is effected. (Furthermore all is rigorous and not just formal, since we have restricted ourselves to $|z| < 1$ wherein convergence is absolute.)

Thus

$$\sum C(n)z^n = \frac{1}{(1-z)(1-z^2)(1-z^3)}, \tag{1}$$

and the generating function for our unknown quantity $C(n)$ is produced. Our number theoretic problem has been translated into a problem about analytic functions, namely, finding the Taylor coefficients of the function $\frac{1}{(1-z)(1-z^2)(1-z^3)}$.

Fine. A well defined analytic problem, but how to solve it? We must resist the temptation to solve this problem by *undoing* the analysis which led to its formulation. Thus the thing *not* to do is expand $\frac{1}{1-z}$, $\frac{1}{1-z^2}, \frac{1}{1-z^3}$ respectively into $\sum z^a, \sum z^{2b}, \sum z^{3c}$ and multiply only to discover that the coefficient is the number of ways of making change for n.

The correct answer, in this case, comes from an *algebraic* technique that we all learned in *calculus*, namely partial fraction. Recall that this leads to terms like $\frac{A}{(1-\alpha z)^k}$ for which we know the expansion explicitly (namely, $\frac{1}{(1-\alpha z)^k}$ is just a constant times the $(k-1)$th derivative of $\frac{1}{(1-\alpha z)} = \sum \alpha^n z^n$).

Carrying out the algebra, then, leads to the partial fractional decomposition which we may arrange in the following form:

$$\frac{1}{(1-z)(1-z^2)(1-z^3)}$$
$$= \frac{1}{6}\frac{1}{(1-z)^3} + \frac{1}{4}\frac{1}{(1-z)^2} + \frac{1}{4}\frac{1}{(1-z^2)} + \frac{1}{3}\frac{1}{(1-z^3)}.$$

Thus, since

$$\frac{1}{(1-z)^2} = \frac{d}{dz}\frac{1}{1-z} = \frac{d}{dz}\sum z^n = \sum (n+1)z^n$$

and

$$\frac{1}{(1-z)^3} = \frac{d}{dz}\frac{1}{2(1-z)^2} = \frac{d}{dz}\sum \frac{n+1}{2}z^n$$
$$= \sum \frac{(n+2)(n+1)}{2}z^n,$$

$$C(n) = \frac{(n+2)(n+1)}{12} + \frac{n+1}{4} + \begin{cases} \frac{1}{4}, & \text{if } n \text{ is even,} \\ \frac{1}{3}, & \text{if } 3|n \end{cases} \tag{2}$$

A somewhat cumbersome formula, but one which can be shortened nicely into

$$C(n) = \left[\frac{n^2}{12} + \frac{n}{2} + 1\right]; \tag{3}$$

where the terms in the brackets mean the greatest integers.

A nice crisp exact formula, but these are rare. Imagine the mess that occurs if the coins were the usual coins of the realm, namely 1, 5, 10, 25, 50, (100?). The right thing to ask for then is an "asymptotic" formula rather than an exact one.

Recall that an *asymptotic* formula $F(n)$ for a function $f(n)$ is one for which $\lim_{n\to\infty} \frac{f(n)}{F(n)} = 1$. In the colorful language of E. Landau, the *relative error* in replacing $f(n)$ by $F(n)$ is *eventually* 0%. At any rate, we write $f(n) \sim F(n)$ when this occurs. One famous such example is Stirling's formula $n! \sim \sqrt{2\pi n}(\frac{n}{e})^n$. (Also note that our result (3) can be weakened to $C(n) \sim \frac{n^2}{12}$.)

So let us assume quite generally that there are coins $a_1, a_2, a_3, \ldots, a_k$, where to avoid trivial congruence considerations we will require that there be no common divisiors other than 1. In this generality we ask for an asymptotic formula for the corresponding $C(n)$. As before we find that the generating function is given by

$$\sum C(n)z^n = \frac{1}{(1-z^{a_1})(1-z^{a_2})\cdots(1-z^{a_k})}. \tag{4}$$

But the next step, explicitly finding the partial fractional decomposition of this function is the *hopeless* task. However, let us simply look for one of the terms in this expansion, the *heaviest* one. Thus at $z = 1$ the denominator has a k-fold zero and so there will be a term $\frac{c}{(1-z)^k}$. All the other zeros are at roots of unity and, because we assumed no common divisiors, all will be of order lower than k.

Thus, although the coefficient of the term $\frac{c}{(1-z)^k}$ is $c\binom{n+k}{k-1}$, the coefficients of all other terms $\frac{a}{(1-\omega z)^3}$ will be $a\omega^j\binom{n+j}{j-1}$. Since all of these j are less than k, the sum total of all of these terms is negligible compared to our one *heavy* term $c\binom{n+k}{k-1}$. In short $C(n) \sim c\binom{n+k}{k-1}$, or even simpler,

$$C(n) \sim c\frac{n^{k-1}}{(k-1)!}.$$

But, what is c? Although we have deftly avoided the necessity of finding all of the other terms, we cannot avoid this one (it's the whole story!). So let us write

$$\frac{1}{(1-z^{a_1})(1-z^{a_2})\cdots(1-z^{a_k})} = \frac{c}{(1-z)^k} + \text{ other terms,}$$

multiply by $(1-z)^k$ to get

$$\frac{1-z}{1-z^{a_1}}\frac{1-z}{1-z^{a_2}}\cdots\frac{1-z}{1-z^{a_k}} = c + (1-z)^k \times \text{ other terms,}$$

and finally let $z \to 1$. By L'Hôpital's rule, for example, $\frac{1-z}{1-z^{a_i}} \to \frac{1}{a_i}$ whereas each of the other terms times $(1-z)^k$ goes to 0. The final result is $c = \frac{1}{a_1a_2\cdots a_k}$, and our final asymptotic formula reads

$$C(n) \sim \frac{n^{k-1}}{a_1a_2\cdots a_k(k-1)!}. \tag{5}$$

Crazy Dice

An ordinary pair of dice consist of two cubes each numbered 1 through 6. When tossed together there are altogether 36 (equally likely) outcomes. Thus the sums go from 2 to 12 with varied numbers of repeats for these possibilities. In terms of our analytic representation, each die is associated with the polynomial $z + z^2 + z^3 + z^4 + z^5 + z^6$. The combined possibilities for the *sums* then are the terms of the *product*

$$
\begin{aligned}
&(z + z^2 + z^3 + z^4 + z^5 + z^6)(z + z^2 + z^3 + z^4 + z^5 + z^6) \\
&\quad = z^2 + 2z^3 + 3z^4 + 4z^5 + 5z^6 + 6z^7 \\
&\qquad + 5z^8 + 4z^9 + 3z^{10} + 2z^{11} + z^{12}
\end{aligned}
$$

The correspondence, for example, says that there are 3 ways for the 10 to show up, the coefficients of z^{10} being 3, etc. The question is Is there any *other* way to number these two cubes with positive integers so as to achieve the very same alternatives?

Analytically, then, the question amounts to the existence of positive integers, $a_1, \cdots, a_6; b_1, \cdots, b_6$, so that

$$
\begin{aligned}
&(z^{a_1} + \cdots + z^{a_6})(z^{b_1} + \cdots + z^{b_6}) \\
&\quad = z^2 + 2z^3 + 3z^4 + \cdots + 3z^{10} + 2z^{11} + z^{12}.
\end{aligned}
$$

These would be the "Crazy Dice" referred to in the title of this section. They look totally different from ordinary dice but they produce exactly the same results!

So, repeating the question, can

$$
\begin{aligned}
&(z^{a_1} + \cdots + z^{a_6})(z^{b_1} + \cdots + z^{b_6}) \\
&\quad = (z + z^2 + z^3 + z^4 + z^5 + z^6) \qquad (6) \\
&\qquad \times (z + z^2 + z^3 + z^4 + z^5 + z^6)?
\end{aligned}
$$

To analyze this possibility, let us factor completely (over the rationals) this right-hand side. Thus $z + z^2 + z^3 + z^4 + z^5 + z^6 = z\frac{1-z^6}{1-z}(1 + z^3) = z(1 + z + z^2)(1 + z)(1 - z + z^2)$. We conclude from (6) that the "a-polynomial" and "b-polynomial" must consist of these factors. Also there are certain side restrictions. The a's and b's are to be *positive* and so a z factor must appear in both polynomials. The a-polynomial must be 6 at $z = 1$ and so the $(1 + z + z^2)(1 + z)$ factor must appear in it, and similarly in the b-polynomial. All that is left to distribute are the two factors of $1 - z + z^2$. If one apiece are given to the a- and b- polynomials, then we get ordinary dice. The only thing left to try is putting both into the a-polynomial.

This works! We obtain finally

$$\sum z^a = z(1 + z + z^2)(1 + z)(1 - z + z^2)^2$$
$$= z + z^2 + z^3 + z^4 + z^5 + z^6 + z^8$$

and

$$\sum z^b = z(1 + z + z^2)(1 + z) = z + 2z^2 + 2z^3 + z^4.$$

Translating back, the crazy dice are 1,3,4,5,6,8 and 1,2,2,3,3,4.

Now we introduce the notion of the *representation function*. So, suppose there is a set A of nonnegative integers and that we wish to express the number of ways in which a given integer n can be written as the sum of two of them. The trouble is that we must decide on conventions. Does order count? Can the two summands be equal? Therefore we introduce *three* representation functions.

$$r(n) = \{\#a, a' \in A, n = a + a'\};$$

So here order counts, and they can be equal;

$$r_+(n) = \{\#a, a' \in A, a \leq a', n = a + a'\},$$

order doesn't count, and they can be equal;

$$r_-(n) = \{\#a, a' \in A, a < a', n = a + a'\},$$

order doesn't count, and they can't be equal. In terms of the generating function for the set A, namely, $A(z) = \sum_{a \in A} z^a$, we can express the generating functions of these representation functions.

The simplest is that of $r(n)$, where obviously

$$\sum r(n)z^n = A^2(z). \tag{7}$$

To deal with $r_-(n)$, we must subtract $A(z^2)$ from $A^2(z)$ to remove the case of $a = a'$ and then divide by 2 to remove the *order*. So here

$$\sum r_-(n)z^n = \frac{1}{2}[A^2(z) - A(z^2)]. \tag{8}$$

Finally for $r_+(n)$, we must add $A(z^2)$ to this result to reinstate the case of $a = a'$, and we obtain

$$\sum r_+(n)z^n = \frac{1}{2}[A^2(z) + A(z^2)]. \tag{9}$$

Can $r(n)$ be "constant?"

Is it possible to design a nontrivial set A, so that, say, $r(n)$ is the same for all n? The answer is NO, for we would have to have $0 \in A$. And the $1 \in A$, else $r_+(1) \neq r_+(0)$. And then $2 \notin A$, else $r_+(2) = 2$. And then $3 \in A$, else $r_+(3) = 0$(whereas $r_+(1) = 1$), then $4 \notin A$, else $r_+(4) = 2$. Continuing in this manner, we find $5 \in A$. But now we are stymied since now $6 = 1 + 5$, $6 = 3 + 3$, and $r_+(6) = 2$.

The suspicion arises, though, that this impossibility may just be a quirk of "small" numbers. Couldn't A be designed so that, except for some misbehavior at the beginning, $r_+(n) = \text{constant}$?

We will analyze this question by using of generating functions. So, using (9), the question reduces to whether there is an infinite set A for which

$$\frac{1}{2}[A^2(z) + A(z^2)] = P(z) + \frac{C}{1 - z}, \tag{10}$$

$P(z)$ is a polynomial.

Answer: No. Just look what happens if we let $z \to (-1)^+$. Clearly $P(z)$ and $\frac{C}{1-z}$ remain bounded, $A^2(z)$ remains nonnegative, and $A(z^2)$ goes to $A(1) = \infty$, a contradiction.

A Splitting Problem

Can we split the nonnegative integers in two sets A and B so that every integer n is expressible in the same number of ways as the sum of two distinct members of A, as it is as the sum of two distinct members of B?

If we experiment a bit, before we get down to business, and begin by placing $0 \in A$, then $1 \in B$, else 1 would be expressible as $a + a'$ but not as $b + b'$. Next $2 \in B$, else 2 would be $a + a'$ but not $b + b'$. Next $3 \in A$, else 3 would *not* be $a + a'$ whereas it is $b + b' = 1 + 2$. Continuing in this manner, we seem to force $A = \{0, 3, 5, 6, 9, \cdots\}$ and $B = \{1, 2, 4, 7, 8, \cdots\}$. But the pattern is not clear, nor is the existence or uniqueness of the desired A, B. We must turn to generating functions. So observe that we are requiring by (8) that

$$\frac{1}{2}[A^2(z) - A(z^2)] = \frac{1}{2}[B^2(z) - B(z^2)]. \tag{11}$$

Also, because of the condition that A, B be a *splitting* of the nonnegatives, we also have the condition that

$$A(z) + B(z) = \frac{1}{1 - z}. \tag{12}$$

From (11) we obtain

$$A^2(z) - B^2(z) = A(z^2) - B(z^2), \tag{13}$$

and so, by (12), we conclude that

$$[A(z) - B(z)] \cdot \frac{1}{1 - z} = A(z^2) - B(z^2),$$

or

$$A(z) - B(z) = (1 - z)[A(z^2) - B(z^2)]. \tag{14}$$

Now this is a relationship that can be iterated. We see that

$$A(z^2) - B(z^2) = (1 - z^2)[A(z^4) - B(z^4)],$$

sot that continuing gives

$$A(z) - B(z) = (1 - z)(1 - z^2)[A(z^4) - B(z^4)].$$

And, if we continue to iterate, we obtain

$$A(z) - B(z) = (1 - z)(1 - z^2) \cdots (1 - z^{2^{n-1}}) \left[A(z^{2^n}) - B(z^{2^n})\right], \tag{15}$$

and so, by letting $n \to \infty$, since $A(0) = 1$, $B(0) = 0$, we deduce that

$$A(z) - B(z) = \prod_{i=0}^{\infty}(1 - z^{2^i}). \tag{16}$$

And this product is easy to "multiply out". Every term z^n occurs uniquely since every n is uniquely the sum of distinct powers of 2. Indeed z^n occurs with $a + 1$ coefficient if n is the sum of an *even* number of distinct powers of 2, and it has $a - 1$ coefficient, otherwise.

We have achieved success! The sets A and B do exist, are unique, and indeed are given by A = Integers, which are the sum of an even number of distinct powers of 2, and B = Integers, which are the sum of an odd number of distinct powers of 2. This is not one of those problems where, after the answer is exposed, one proclaims, "oh, of course." It isn't really trivial, even in retrospect, why the A and B have the same $r_-(n)$, or for that matter, to what this common $r_-(n)$ is equal. (See enclosure where it is proved that $r_-(2^{2k+1} - 1) = 0$.)

A = Integers with an *even* number of 1's in radix 2. Then and only then

$$\overbrace{111\cdots 1}^{2k+1} = 2^{2k+1} - 1$$

is *not* the sum of two distinct A's.

PROOF. A sum of two A's, with *no* carries has an even number of 1's (so it won't give $\overbrace{111\cdots 1}^{\text{odd}}$), else look at the *first* carry. This gives a 0 digit so, again, it's not $11\cdots 1$.

So $r_-(2^{2k+1} - 1) = 0$. We must now show that all other n have a representation as the sum of two numbers whose numbers of 1 digits are of like parity. First of all if n contains zk 1's then it is the sum of the first k and the second k. Secondly if n contains $2k + 1$ 1's but also a 0 digit then it is structured as $\underbrace{111\cdots}_{m}\circ A$ where A contains $2k + 1 - m$ 1's and, say, is of total length L then it can be expressed as $\underbrace{111\cdots 1}_{m-1}\circ\underbrace{00\cdots 00}_{2}$ plus $1A$ and these two numbers have respectively m 1's and $2k + 2 - m$ 1's. These are again of like parity so Q.E.D.

An Identity of Euler's

Consider expressing n as the sum of distinct positive integers, i.e., where repeats are not allowed. (So For $n = 6$, we have the expression $1 + 2 + 3$ and also $2 + 4$, $1 + 5$, and just plain 6 alone.)

Also consider expressing n as the sum of positive odd numbers, but this time where repeats *are* allowed. (So for $n = 6$, we get $1 + 5$, $3 + 3$, $1 + 1 + 1 + 3$, $1 + 1 + 1 + 1 + 1 + 1$.) In both cases we obtained four expressions for 6, and a theorem of Euler's says that this is no coincidence, that is, it says the following:

Theorem. *The number of ways of expressing n as the sum of distinct positive integers equals the number of ways of expressing n as the sum of (not necessarily distinct) odd positive integers.*

To prove this theorem we produce two generating functions. The latter is exactly the "coin changing" function where the coins have the denominations 1, 3, 5, 7, ... This generating function is given by

$$\frac{1}{(1-z)(1-z^3)(1-z^5)\cdots}. \tag{17}$$

The other generating function is not of the coin changing variety because of the distinctness condition. A moment's thought, however, shows that this generating function is given as the product of $1 + z$, $1 + z^2$, $1 + z^3$, $\cdots$ for, when these are multiplied out, each z^k factor occurs at most once. In short, the other generating function is

$$(1+z)(1+z^2)(1+z^3)\cdots. \tag{18}$$

Euler's theorem in its analytic form is just the identity

$$\frac{1}{(1-z)(1-z^3)(1-z^5)\cdots} = (1+z)(1+z^2)(1+z^3)\cdots \quad \text{throughout } |z| < 1. \tag{19}$$

Another way of writing (19) is

$$(1-z)(1-z^3)(1-z^5)\cdots(1+z)(1+z^2)(1+z^3)\cdots = 1 \tag{20}$$

which is the provocative assertion that, when this product is multiplied out, all of the terms (aside from the 1) *cancel* each other!

To prove (2) multiply the $1 - z$ by the $1 + z$ (to get $1 - z^2$) and do the same with $1 - z^3$ by $1 + z^3$, etc. This gives the new factors $1 - z^2, 1 - z^6, 1 - z^{10}, \cdots$ and leaves untouched the old factors $1 + z^2, 1 + z^4, 1 + z^6, \cdots$. These rearrangements are justified by absolute convergence, and so we see that the product in (20), call it $P(z)$, is equal to

$$(1 - z^2)(1 - z^6)(1 - z^{10}) \cdots (1 + z^2)(1 + z^4) \cdots$$

which just happens to be $P(z^2)$! So $P(z) = P(z^2)$ which of course means that there can't be any terms az^k, $a \neq 0$, $k \neq 0$, in the expansion of $P(z)$, i.e., $P(z)$ is just its constant term 1, as asserted.

Marks on a Ruler

Suppose that a 6" ruler is marked as usual at 0, 1, 2, 3, 4, 5, 6. Using this ruler we may of course measure any integral length from 1 through 6. But we don't need all of these markings to accomplish these measurements. Thus we can remove the 2, 3, and 5, and the marks at 0, 1, 4, 6 are sufficient. (The 2 can be measured between 4 and 6, the 3 can be gotten between 1 and 4, and the 5 between 1 and 6.) Since $\binom{4}{2} = 6$, this is a "perfect" situation. The question suggests itself then, are there any *larger* perfect values? In short, can there be integers $a_1 < a_2 < \cdots < a_n$ such that the differences $a_i - a_j$, $i > j$, take on *all* the values $1, 2, 3, \ldots, \binom{n}{2}$?

If we introduce the usual generating function $A(z) = \sum_{i=1}^{n} z^{a_i}$, then the differences are exposed, not when we square $A(z)$, but when we multiply $A(z)$ by $A(\frac{1}{z})$. Thus $A(z) \cdot A(\frac{1}{z}) = \sum_{i,j=1}^{n} z^{a_i - a_j}$ and if we split this (double) sum as $i > j$, $i = j$, and $i < j$, we obtain

$$A(z) \cdot A\left(\frac{1}{z}\right) = \sum_{i,j=1_{i>j}}^{n} z^{a_i - a_j} + n + \sum_{i,j=1_{i<j}}^{n} z^{a_i - a_j}.$$

Our "perfect ruler," by hypothesis, then requires that the first sum be equal to $\sum_{k=1}^{N} z^k$, $N = \binom{n}{2}$, and since the last sum is the same as first, with $\frac{1}{z}$ replacing z, our equation takes the simple form

$$A(z) \cdot A\left(\frac{1}{z}\right) = \sum_{k=-N}^{N} z^k + n - 1, \cdot N = \binom{n}{2},$$

or, summing this geometric series,

$$A(z) \cdot A\left(\frac{1}{z}\right) = \frac{z^{N+1} - z^N}{z - 1} + n - 1, \quad N = \binom{n}{2}. \tag{21}$$

In search of a contradiction, we let z lie on the unit circle $z = e^{i\theta}$, so that the left side of (21) becomes simply $|A(e^{i\theta})|^2$, whereas the right-hand side is

$$\frac{z^{N+\frac{1}{2}} - z^{-(N+\frac{1}{2})}}{z^{\frac{1}{2}} - z^{-\frac{1}{2}}} + n - 1 = \frac{\sin(N + \frac{1}{2})\theta}{\sin \frac{1}{2}\theta} + n - 1$$

and (21) reduces to

$$\left|A(e^{i\theta})\right|^2 = \frac{\sin \frac{n^2-n+1}{2}\theta}{\sin \frac{1}{2}\theta} + n - 1. \tag{22}$$

A contradiction will occur, then, if we pick a θ which makes

$$\frac{\sin \frac{n^2-n+1}{2}\theta}{\sin \frac{1}{2}\theta} < -(n - 1). \tag{23}$$

(And we had better assume that $n \geq 5$, since we *saw* the perfect ruler for $n = 4$.)

A good choice, then, is to make $\sin \frac{n^2-n+1}{2}\theta = -1$, for example by picking $\theta = \frac{3\pi}{n^2-n+1}$. In that case $\sin \frac{\theta}{2} < \frac{\theta}{2}$, $\frac{1}{\sin \frac{\theta}{2}} > \frac{2}{\theta}$, $-\frac{1}{\sin \frac{\theta}{2}} < -\frac{2}{\theta} = -\frac{2n^2-2n+2}{3\pi}$. and so the requirement (23) follows from $-\frac{2n^2-2n+2}{3\pi} < -(n - 1)$ or $2n^2 - 2n + 2 > 3\pi(n - 3)$. But $2n^2 - 2n + 2 - 3\pi(n - 1) > 2n^2 - 2n + 2 - 10(n - 1) = 2(n - 3)^2 - 6 \geq 2 \cdot 2^2 - 6 = 2$, for $n \geq 5$. There are no perfect rulers!

Dissection into Arithmetic Progressions

It is easy enough to split the nonnegative integers into arithmetic progressions. For example they split into the *evens* and the *odd* or into the progressions $2n$, $4n + 1$, $4n + 3$. Indeed there are many other ways, but all seem to require at least two of the progressions to have same common difference (the evens and odds both have 2 as a common difference and the $4n + 1$ and $4n + 3$ both have 4). So the question arises Can the positive integers be split into at least two arithmetic progressions any two of which have a distinct common difference?

Of course we look to generating functions for the answer. The progression $an + b$, $n = 1, 2, 3, \ldots$ will be associated with the function $\sum_{n=1}^{\infty} z^{an+b}$. Thus the dissection into evens and odds corresponds to the identity $\sum_{n=0}^{\infty} z^n = \sum_{n=0}^{\infty} z^{2n} + \sum_{n=0}^{\infty} z^{2n+1}$, and the dirssection into $2n$, $4n + 1$, $4n + 3$ corresponds to $\sum_{n=0}^{\infty} z^n = \sum_{n=0}^{\infty} z^{2n} + \sum_{n=0}^{\infty} z^{4n+1} + \sum_{n=0}^{\infty} z^{4n+3}$, etc. Since each of these series is geometric, we can express their sums by $\sum_{n=1}^{\infty} z^{an+b} = \frac{z^b}{1-z^a}$. Our question then is exactly whether there can be an identity

$$\frac{1}{1-z} = \frac{z^{b_1}}{1-z^{a_1}} + \frac{z^{b_2}}{1-z^{a_2}} + \cdots + \frac{z^{b_k}}{1-z^{a_k}},$$
$$1 < a_1 < a_2 \cdots < a_k. \tag{24}$$

Well, just as the experiment suggested, there *cannot* be such a dissection, (24) is impossible. To see that (24) does, indeed, lead to a contradiction, all we need do is let $z \to e^{\frac{2\pi i}{a_k}}$ and observe that then all of the terms in (24) approach finite limits except the last term $\frac{z^{b_k}}{1-z^{a_k}}$ which approaches ∞. Hopefully, then, this chapter has helped take the *sting* out of the preposterous notion of using analysis in number theory.

Problems for Chapter I

1. Procduce a set A such that $r(n) > 0$ for all n in $1 \leq n \leq N$, but with $|A| \leq \sqrt{4N+1}$.

2. Show that evert set satisfying the conditions of (1) must have $|A| \leq \sqrt{N}$.

3. Show directly, with no knowledge of Stirling's formula, that $n! > (\frac{n}{e})^n$.

II

The Partition Function

One of the simplest, most natural, questions one can ask in arithmetic is how to determine the number of ways of *breaking up* a given integer. That is, we ask about a positive integer n In how many ways can it be written as $a + b + c + \cdots$ where $a, b, c, \ldots$ are positive integers? It turns out that there are two distinct questions here, depending on whether we elect to count the *order* of the summands. If we do choose to let the order count, then the problem becomes *too* simple. The answer is just 2^{n-1} and the proof is just induction. Things are incredibly different and more complicated if order is not counted!

In this case the number of breakups or "partitions" is 1 for $n = 1$, 2 for $n = 2$, 3 for $n = 3$, 5 for $n = 4$, 7 for $n = 5$, which 5 has the representations $1 + 1 + 1 + 1 + 1, 2 + 1 + 1 + 1, 3 + 1 + 1, 4 + 1,$ $5,\ 3 + 2,\ 2 + 2 + 1$, and no others. *Remember* such expressions as $1 + 1 + 2 + 1$ are not considered different. The table can be extended further of course but no apparent pattern emerges. There is a famous story concerning the search for some kind of pattern in this table. This is told of Major MacMahon who kept a list of these partition numbers arranged one under another up into the hundreds. It suddenly occurred to him that, viewed from a distance, the outline of the digits seemed to form a parabola! Thus the number of digits in $p(n)$, the number of partitions of n, is around $C\sqrt{n}$, or $p(n)$ itself is very roughly $e^{\alpha\sqrt{n}}$. The first crude assessment of $p(n)$!

Among other things, however, this does tell us not to expect any simple answers. Indeed later research showed that the true asymptotic formula for $p(n)$ is $\frac{e^{\pi}\sqrt{2n/3}}{4\sqrt{3}n}$, certainly not a formula to be guessed!

Now we turn to the analytic number theory derivation of this asymptotic formula.

The Generating Function

To put into sharp focus the fact that order does not count, we may view $p(n)$ as the number of representations of n as a sum of 1's and 2's and 3's . . . , etc. But this is just the "change making" problem where coins come in all denominations. The analysis in that problem extends verbatim to this one, even though we now have an infinite number of coins, So we obtain

$$\sum_{n=0}^{\infty} p(n)z^n = \sum_{k=1}^{\infty} \frac{1}{1-z^k} \tag{1}$$

valid for $|z| < 1$, where we understand that $p(0) = 1$.

Having thus obtained the generating function, we turn to the second stage of attack, investigating the function. This is always the tricky (creative?) part of the process. We know pretty well what kind of information we desire about $p(n)$: an estimate of its growth, perhaps even an asymptotic formula if we are lucky. But we don't know exactly how this translates to the generating function. To grasp the connection between the generating function and its coefficients, then, seems to be the paramount step. How does one go from one to the other? Mainly how does one go from a function to its coefficients?

It is here that complex numbers really play their most important role. The point is that there are formulas (for said coefficients). Thus we learned in calculus that, if $f(z) = \sum a_n z^n$, then $a_n = \frac{f^{(n)}(0)}{n!}$, expressing the desired coefficients in terms of high derivatives of the function. But this a *terrible* way of getting at the thing. Except for rare "made up" examples there is very little hope of obtaining the nth derivative of a given function and even *estimating* these derivatives is not a task with very good prospects. Face it, the calculus approach is a flop.

Cauchy's theorem gives a different and more promising approach. Thus, again with $f(z) = \sum a_n z^n$, this time we have the formula

$a_n = \frac{1}{2\pi i} \int_C \frac{f(z)}{z^{n+1}} dz$, an integral rather than a differential operator! Surely this is a more secure approach, because integral operators are *bounded*, and differential operators are not. The price we pay is that of passing to the complex numbers for our z's. Not a bad price, is it?

So let us get under way, but armed with the knowledge that the valuable information about $f(z)$ will help in getting a good approximation to $\int_C \frac{f(z)}{z^{n+1}} dz$. But a glance at the potentially explosive $\frac{1}{z^{n+1}}$ shows us that C had better stay as far away from the origin as it can, i.e., it must hug the unit circle. Again, a look at our generating function $\sum p(n)z^n$ shows that it's biggest when z is positive (since the coefficients are themselves positive). All in all the, we see that we should seek approximations to our generating function which are good for $|z|$ near 1 with special importance attached to those z's which are near $+1$.

The Approximation

Starting with (1), $F(z) = \prod_{k=1}^{\infty} \frac{1}{1-z^k}$, and taking logarithms, we obtain

$$\log F(z) = \sum_{k=1}^{\infty} \log \frac{1}{1-z^k} = \sum_{k=1}^{\infty} \sum_{j=1}^{\infty} \frac{z^{kj}}{j}$$

$$= \sum_{j=1}^{\infty} \frac{1}{j} \sum_{k=1}^{\infty} z^{jk} = \sum_{j=1}^{\infty} \frac{1}{j} \frac{z^j}{1-z^j}. \tag{2}$$

Now write $z = e^{-w}$ so that $\Re w > 0$ and obtain $\log F(e^{-w}) = \sum_{k=1}^{\infty} \frac{1}{k} \frac{1}{e^{kw}-1}$. Thus noticing that the expansion of $\frac{1}{e^x-1}$ begins with $\frac{1}{x} - \frac{1}{2} + c_1 x + \cdots$ or equivalently (near 0) $\frac{1}{x} - \frac{e^{-x}}{2} + cx + \cdots$, we rewrite this as

$$\log F(e^{-w}) = \sum \frac{1}{k} \left(\frac{1}{kw} - \frac{e^{-kw}}{2} \right)$$

$$+ \sum \frac{1}{k} \left(\frac{1}{e^{kw}-1} - \frac{1}{kw} + \frac{e^{-kw}}{2} \right) \tag{3}$$

$$= \frac{\pi^2}{6w} + \frac{1}{2}\log(1 - e^{-w})$$
$$+ \sum \frac{1}{k}\left(\frac{1}{e^{kw} - 1} - \frac{1}{kw} + \frac{e^{-kw}}{2}\right).$$

The form of this series is very suggestive. Indeed we recognize any series $\sum \frac{1}{k} A(kw) = \sum \frac{A(kw)}{kw} w$ as a Riemann sum, approximating the Riemann integral $\int_0^\infty \frac{A(t)}{t} dt$ for small *positive* w. It should come as no surprise then, that such series are estimated rather accurately. So let us review the "Riemann sum story".

Riemann Sums

Suppose that $\phi(x)$ is a positive decreasing function on $(0, \infty)$ and that $h > 0$. The Riemann sum $\sum_{k=1}^\infty \phi(kh)h$ is clearly equal to the area of the union of rectangles and so is bounded by the area under $y = \phi(x)$. Hence $\sum_{k=1}^\infty \phi(kh)h < \int_0^\infty \phi(x)dx$. On the other hand, the series $\sum_{k=0}^\infty \phi(kh)h$ can be contrued as the area of this union of these rectangles and, as such, exceeds the area under $y = \phi(x)$. So this time we obtain $\sum_{k=0}^\infty \phi(kh)h > \int_0^\infty \phi(x)dx$.

Combining these two inequalities tells us that the Riemann sum lies within $h \cdot \phi(0)$ of the Riemann integral. This is all very nice and rather accurate but it refers only to decreasing functions. However, we may easily remedy this restriction by subtracting two such functions. Thereby we obtain

$$\sum_{k=1}^{\infty}[\phi(kh) - \psi(ki)]h - \int_0^\infty [\phi(x) - \psi(x)] \ll h[\phi(0) + \psi(0)].$$

Calling $\phi(x) - \psi(x) = F(x)$ and then observing that $\phi(0) + \psi(0)$ is the total variation V of $F(x)$ we have the rather general result

$$\sum_{k=1}^{\infty} F(kh)h - \int_0^\infty F(x) \ll h \cdot V(F). \tag{4}$$

To be sure, we have proven this result only for real functions but in fact it follows for complex ones, by merely applying it to the real and imaginary parts.

To modify this result to fit our situation, let us write $w = he^{i\theta}$, $h > 0, -\pi/2 < \theta < \pi/2$, and conclude from(4) that

$$\sum_{k=1}^{\infty} F(khe^{i\theta})h - \int_0^{\infty} F(xe^{i\theta}dx) \ll h \cdot V_\theta(F)$$

(V_θ is the variation along the ray of argument θ), so that

$$\sum_{k=1}^{\infty} F(kw)w - \int_0^{\infty} F(xe^{i\theta})d(xe^{i\theta}) \ll w \cdot V_\theta(F).$$

Furthermore, in our case of an *analytic* F, this integral is actually independent of θ. (Simply apply Cauchy's theorem and observe that at ∞ F falls off like $\frac{1}{x^2}$). We also may use the formula $V_\theta(F) = \int_0^\infty |f'(xe^{i\theta})|dx$ and finally deduce that

$$\sum_{k=1}^{\infty} F(kw)w - \int_0^{\infty} F(x)dx \ll w \int_0^{\infty} |F'(xe^{i\theta})|dx.$$

Later on we show that

$$\int_0^{\infty} \left(\frac{1}{e^x - 1} - \frac{1}{x} + \frac{e^{-x}}{2} \right) \frac{dx}{x} = \log \frac{1}{\sqrt{2\pi}}, \tag{5}$$

and right now we may note that the (complicated) function

$$F'(xe^{i\theta}) = \frac{-e^{xe^{i\theta}}}{xe^{i\theta}(e^{xe^{i\theta}} - 1)^2} + \frac{2}{x^3 e^{2i\theta}} - \frac{e^{-xe^{i\theta}}}{2x} - \frac{e^{-xe^{i\theta}}}{2x^2}$$

is *uniformly* bounded by $\frac{M}{(z+1)^2}$ in any wedge $|\theta| < c < \pi/2(m + M(c))$, so that we obtain

$$\sum_{k=1}^{\infty} \left(\frac{1}{e^k w - 1} - \frac{1}{kw} + \frac{e^{-kw}}{2} \right) - \log \frac{1}{\sqrt{2\pi}} \ll Mw \tag{6}$$

throughout $|\arg w| < c < \pi/2$.

The Approximation. We have prepared the way for the useful approximation to our generating function. All we need to do is combine

(1), (3), and (6), replace w by $\log \frac{1}{z}$, and exponentiate. The result is

$$\prod_{k=1}^{\infty} \frac{1}{1-z^k}$$

$$= \sqrt{\frac{1-z}{2\pi}}\, exp\left(\frac{\pi^2}{6 \log \frac{1}{z}}\right)[1 + O(1-z)]$$

$$\text{in} \quad \frac{|1-z|}{1-|z|} \leq c.$$

But we perform one more "neatening" operation. Thus $\log \frac{1}{z}$ is an eyesore! It isn't at all analytic in the unit disc, we must replace it (before anything good can result). So note that, near 1, $\log \frac{1}{z} = (1-z) + \frac{(1-z)^2}{2} + \frac{(1-z)^3}{3} + \cdots = 2\frac{1-z}{1+z} + O(1-z)^3$, or $\frac{1}{\log \frac{1}{z}} = \frac{1}{2}\frac{1+z}{1-z} + O(1-z)$. Finally then,

$$\prod_{k=1}^{\infty} \frac{1}{1-z^k}$$

$$= \sqrt{\frac{1-z}{2\pi}}\, exp\left(\frac{\pi^2}{12}\frac{1+z}{1-z}\right)[1 + O(1-z)] \tag{7}$$

$$\text{in} \quad \frac{|1-z|}{1-|z|} / lec.$$

This is our basic approximation. It is good near $z = 1$, which we have decided is the most important locale. Here we see that we can replace our generating function by the *elementary* function $\sqrt{\frac{1-z}{2\pi}}\, exp\left(\frac{\pi^2}{12}\frac{1+z}{1-z}\right)$ whose coefficients should then prove amenable.

However, (7) is really of no use away from $z = 1$, and, since Cauchy's theorem requires values of z all along a closed loop surrounding 0, we see that something else must be supplied. Indeed we will show that, away from 1, everything is negligible by comparison.

To see this, let us return to (2) and conclude that

$$\log F(z) - \frac{1}{1-z} \ll \sum_{j=2}^{\infty} \frac{1}{j} \frac{|z|^j}{1-|z|^j} \ll \frac{1}{1-|z|} \sum_{j=2}^{\infty} \frac{1}{j} \frac{1}{j}$$

$$= \frac{1}{1-|z|} \left(\frac{\pi^2}{6} - 1 \right),$$

or

$$F(z) \ll \exp \left(\frac{1}{|1-z|} + \left(\frac{\pi^2}{6} - 1 \right) \frac{1}{1-|z|} \right), \tag{8}$$

an estimate which is just what we need. It shows that, away from 1, where $\frac{1}{|1-z|}$ is smaller than $\frac{1}{1-|z|}$, $F(z)$ is rather small.

Thus, for example, we obtain

$$F(z) \ll \exp \frac{1}{|1-z|} \quad \text{when} \quad \frac{|1-z|}{1-|z|} \geq 3. \tag{9}$$

Also, in this same region, setting

$$\phi(z) = \sqrt{\frac{1-z}{2\pi}} \exp \left(\frac{\pi^2}{12} \frac{1+z}{1-z} \right) = \sum q(n) z^n, \tag{10}$$

$$\phi(z) \ll \sqrt{\frac{2}{2\pi}}\, exp \left(\frac{\pi^2}{12} \frac{2}{1-z} \right) \ll exp \left(\frac{\pi^2}{12} \frac{2}{3(1-|z|)} \right)$$

so that

$$\phi(z) \ll exp \frac{1}{1-|z|}, \tag{11}$$

when $\frac{|1-z|}{1-|z|} \geq 3$.

The Cauchy Integral. Armed with these preparations and the feeling that the coefficients of the elementary function $\phi(z)$ are accessible, we launch our major Cauchy integral attack. So, to commence the firing, we write

$$p(n) - q(n) = \frac{1}{2\pi i} \int_C \frac{F(z) - \phi(z)}{z^{n+1}} dz \tag{12}$$

and we try C a circle near the unit circle, i.e.,

$$C \quad \text{is} \quad |z| = r, r < 1. \tag{13}$$

Next we break up C as dictated by our consideration of $\frac{|1-z|}{1-|z||}$, namely, into

$$A \quad \text{is the arc} \quad |x| = r, \frac{|1-z|}{1-|z|} \leq 3,$$

and (14)

$$B \quad \text{is the arc} \quad |x| = r, \frac{|1-z|}{1-|z|} \geq 3$$

So,

$$p(n) - q(n) \tag{15}$$
$$= \frac{1}{2\pi i}\int_A \frac{F(z)-\phi(z)}{z^{n+1}}dz + \frac{1}{2\pi i}\int_B \frac{F(z)-\phi(z)}{z^{n+1}}dz,$$

and if we use (7) on this first integral and (9), (11) on this second integral we derive the following estimates:

$$\frac{1}{2\pi i}\int_A \frac{F(z)-\phi(z)}{z^{n+1}}dz$$
$$\ll \frac{M'}{r^{n+1}}(1-r)^{3/2}\exp\left(\frac{\pi^2}{6}\frac{1}{1-r}\right) \times \text{the length of } A.$$

(M' is the implied constant in the O of (7) when $c = 3$).

As for the length of A, elementary geometry gives the formula

$$4r \arcsin \frac{\sqrt{2}(1-r)}{\sqrt{r}}$$

and this is easily seen to be $O(1-r)$. We finally obtain, then,

$$\frac{1}{2\pi i}\int_A \frac{F(z)-\phi(z)}{z^{n+1}}dz$$
$$\ll M\frac{(1-r)^{5/2}.}{r^n}\exp\left(\frac{\pi^2}{6}\frac{1}{1-r}\right), \tag{16}$$

where M is an absolute constant.

For the second integral,

$$\frac{1}{2\pi i}\int_B \frac{F(z)-\phi(z)}{z^{n+1}}\,dz \ll \frac{1}{2\pi r^{n+1}}\cdot 2\exp\frac{1}{1-r}\cdot 2\pi r$$
$$= \frac{2}{r^n}\exp\left(\frac{1}{1-r}\right).$$

And this is even smaller than our previous estimate. So combining the two gives, by (15),

$$p(n)-q(n) \ll M\frac{(1-r)^{5/2}}{r^n}\exp\left(\frac{\pi^2}{6}\frac{1}{1-r}\right). \tag{17}$$

But what is r? Answer: anything we please (as long as $0<r<1$)! We are masters of the choice, and so we attempt to minimize the right-hand side. The exact minimum is too complicated but the approximate one occurs when $\frac{1}{e^{n(r-1)}}\exp\left(\frac{\pi^2}{6}\frac{1}{1-r}\right)$ is minimized and this occurs when $\frac{\pi^2}{6}\frac{1}{1-r}=n(1-r)$, i.e., $r=1-\frac{\pi}{\sqrt{6n}}$. So we choose this r and, by so doing, we obtain, from (17), the bound

$$p(n)=q(n)+O\left(n^{-5/4}e^{\pi\sqrt{2n/3}}\right). \tag{18}$$

The Coefficients of $q(n)$

The elementary function $\phi(z)$ has a rather pleasant definite integral representation which will then lead to a handy expression for the $q(n)$.

If we simply begin with the well-known identity

$$\int_{-\infty}^{\infty} e^{-t^2}\,dt=\sqrt{\pi}$$

and make a linear change of variables,

$$\int_{-\infty}^{\infty} e^{-(at-b)^2}\,dt=\frac{\sqrt{\pi}}{a},$$

or

$$\int_{-\infty}^{\infty} e^{-a^2t^2} e^{+2abt}\, dt = \frac{\sqrt{\pi}}{a} e^{b^2}.$$

Thus if we set $b^2 = \frac{\pi^2}{6}\frac{1}{1-z}$ and $a^2 = 1 - z$ (thinking of z as real, for now), we obtain

$$\int_{-\infty}^{\infty} e^{zt^2} e^{+\pi\sqrt{\frac{2}{3}}t - t^2}\, dt = \frac{\sqrt{\pi}}{1-z} \exp\left(\frac{\pi}{6}\frac{1}{1-z}\right),$$

which gives, finally,

$$\phi(z) = \frac{e^{-\pi^2/12}}{\pi\sqrt{2}}(1-z)\int_{-\infty}^{\infty} e^{zt^2} e^{\pi\sqrt{\frac{2}{3}}t - t^2}\, dt. \tag{19}$$

Equating coefficients therefore results in

$$q(n) = \frac{e^{-\pi^2/12}}{\pi\sqrt{2}} \int_{-\infty}^{\infty} \left[\frac{t^{2n}}{n!} - \frac{t^{2n-2}}{(n-1)!}\right] e^{\pi\sqrt{\frac{2}{3}} - t^2}\, dt \tag{20}$$

the "formula" for $q(n)$ from which we can obtain asymptotics.

Reasoning that the maximum of the integrand occurs near $t = \sqrt{n}$ we change variables by $t = s + \sqrt{n}$, and thereby obtain

$$q(n) = C_n \int_{-\infty}^{\infty} K_n(s) 2s e^{-2\left(s - \frac{\pi}{2\sqrt{6}}\right)^2}\, ds,$$

where

$$C_n = \frac{e^{\pi\sqrt{2n/3}}}{\pi\sqrt{2n}} \frac{n^{n+\frac{1}{2}}}{e^n n!},$$

$$K_n(s) = \frac{1 + \frac{s}{2\sqrt{n}}}{\left(1 + \frac{s}{\sqrt{n}}\right)^2} \left[\left(1 + \frac{s}{\sqrt{n}}\right) e^{\frac{-s}{\sqrt{n}} + \frac{s^2}{2n}}\right].$$

Since $K_n(s) \to 1$, we see, at least is formally, that the above integral approaches

$$\int_{-\infty}^{\infty} 2s e^{-2\left(s - \frac{\pi}{2\sqrt{6}}\right)^2}\, ds = \int_{-\infty}^{\infty} \left(u + \frac{\pi}{2\sqrt{3}}\right) e^{-u^2}\, du,$$

where we have set $s = \frac{u}{\sqrt{2}} + \frac{\pi}{2\sqrt{6}}$. Furthermore, since ue^{-u^2} is odd, it is equal to $\frac{\pi}{2\sqrt{3}} \int_{-\infty}^{\infty} e^{-u^2} du = \frac{\pi\sqrt{\pi}}{2\sqrt{3}}$. Thus (21) formally becomes

$$q(n) \sim \frac{e^{\pi\sqrt{2n/3}}}{4\sqrt{3}n} \frac{\sqrt{2\pi n}n^n}{e^n n!}. \tag{22}$$

And score another one for Stirling's formula, which in turn gives

$$q(n) \sim \frac{e^{\pi\sqrt{2n/3}}}{4\sqrt{3}n}, \tag{23}$$

and our earlier estimate (18) allows us thereby to conclude that

$$p(n) \sim \frac{e^{\pi\sqrt{2n/3}}}{4\sqrt{3}n}. \tag{24}$$

Success! We have determined the asymptotic formula for $p(n)$! Well, almost. We still have two debts outstanding. We must justify our formal passage to the limit in (21), and we must also prove our evaluation (5). So first we observe that xe^{-x} is maximized at $x = 1$, so we deduce that

$$\left(1 + \frac{s}{\sqrt{n}}\right) e^{\frac{-s}{\sqrt{n}}} \leq 1 \tag{25}$$

(using $x = (1 + \frac{s}{\sqrt{n}})$) and also

$$\left|1 + \frac{s}{\sqrt{n}}\right| e^{\frac{-s}{\sqrt{n}}} \leq e^{\frac{s^2}{2n}} \tag{26}$$

(using $x = (1 + \frac{s}{\sqrt{n}})^2$).

Thus using (25) for positive s, by (21),

$$K_n(s) \leq e^{s^2} \quad \text{for} \quad s \geq 0,$$

and using (26) for negative s gives us

$$\begin{aligned}|K_n(s)| &\le (1-s)e^{s^2-\frac{2s}{\sqrt{n}}}\left(\left|1+\frac{s}{\sqrt{n}}\right|e^{\frac{-s}{\sqrt{n}}}\right)^{2n-2}\\ &\le (1-s)e^{s^2-\frac{2s}{\sqrt{n}}}e^{\frac{n-1}{n}s^2}\\ &= (1-s)e^{2s^2+1-\left(\frac{s}{\sqrt{n}}\right)^2} \le (1-s)e^{2s^2+1},\end{aligned}$$

or

$$|K_n(s)| \le (1-s)e^{2s^2+1} \quad \text{for} \quad s<0. \tag{28}$$

Thus (27) and (28) give the bound for our integral in (21) of

$$2se^{s^2-2\left(s-\frac{\pi}{2\sqrt{6}}\right)^2} \quad \text{for} \quad s \ge 0,$$

and

$$2s(s-1)e^{1+\frac{\pi}{\sqrt{6}}s} \quad \text{for} \quad s<0.$$

This bound, integrable over $(-\infty,\infty)$, gives us the required dominated convergence, and the passage to the limit is indeed justified.

Finally we give the following:

Evaluation of our Integral (5). To achieve this let us first note that as $N \to \infty$ our integral is the limit of the integral

$$\int_{-\infty}^{\infty}(1-e^{-Nx})\left(\frac{1}{e^x-1}-\frac{1}{x}+\frac{e^{-x}}{2}\right)\frac{dx}{x}$$

(by dominated convergence, e.g.). But this integral can be split into

$$\begin{aligned}&\int_0^{\infty}(1-e^{-Nx})\left(\frac{1}{e^x-1}-\frac{1}{x}\right)\frac{dx}{x}+\int_0^{\infty}(1-e^{-Nx})\frac{e^{-x}}{2x}dx\\ &=\sum_{k=1}^{N}\int_0^{\infty}e^{-kx}\frac{1+x-e^x}{x^2}dx+\frac{1}{2}\int_0^{\infty}\frac{e^{-x}-e^{-(n+1)x}}{x}dx.\end{aligned}$$

Next note that

$$\frac{1+x-e^x}{x^2} = -\int_0^1 te^{(1-t)x}dt$$

and

$$\frac{e^{-x} - e^{-(n+1)x}}{x} = \int_1^{N+1} e^{-sx} ds.$$

Hence, by Fubini, we may interchange and obtain, for our expression, the elementary sum

$$-\sum_{k=1}^{N} \int_0^1 \frac{t}{k+t+1} dt + \frac{1}{2} \int_1^{N+1} \frac{ds}{s}$$

$$= \sum_{k=1}^{N} [(k-1) \log \frac{k}{k-1} - 1] + \frac{1}{2} \log(N+1)$$

$$= \sum_{k=1}^{N} (k-1) \log k - (k-1) \log(k-1) - N$$

$$+ \frac{1}{2} \log(N+1)$$

$$= N \log N - \log N - \log(N-1) - \cdots - \log 1 - N$$

$$+ \frac{1}{2} \log(N+1)$$

$$= N \log N - \log N! - N + \frac{1}{2} \log(N+1).$$

What luck! This is equal to $\log \frac{\sqrt{N+1}(N/e)^N}{N!}$ and so, by Stirling's formula, indeed approaches $\log \frac{1}{\sqrt{2\pi}}$.

(Stirling's formula was used *twice* and hence needn't have been used at all! Thus we ended up not needing the fact that $C = \sqrt{2\pi}$ in the formula $n! \sim C\sqrt{n}(n/e)^n$ since the c cancels against a c in the denominator. The $n!$ formula with c instead of $\sqrt{2\pi}$ is a much simpler result.)

Problems for Chapter II

1. Explain the observation that MacMahon made of a *parabola* when he viewed the list of the (decimal expansions) of the partition function.

2. Prove the "simple" fact that, if order counts (e.g., $2+5$ is considered a different partition of 7 than $5+2$), then the total number of partitions on n would be 2^{n-1}.

3. Explain the approximation "near 1" of $\log\frac{1}{x}$ as $2\frac{1-z}{1+z}+O(1-z)^3$. Why does this lead to

$$\frac{1}{\log\frac{1}{z}}=\frac{1}{2}\frac{1+z}{1-z}+O(1-z)?$$

4. Why is the Riemann sum such a good approximation to the integral when the function is monotone and the increments are equal?

III

The Erdös–Fuchs Theorem

There has always been some fascination with the possibility of *near* constancy of representation function $r_i(n)$ (of I(7), (8), and (9)). In Chapter I we treated the case of $r_+(n)$ and showed that this could not eventually be constant. The fact that $r(n)$ cannot be constant for an infinite set is really trivial since $r(n)$ is *odd* for $n = 2a$, $a \in A$, and even otherwise. The case of $r - (n)$ is more difficult, and we will treat it in this chapter as an introduction to the analysis in the Erdös–Fuchs theorem.

The Erdös–Fuchs theorem involves the question of just how nearly constant $r(n)$ can be *on average*. Historically this all began with the set A = the perfect squares and the observation that then $\frac{r(0)+r(1)+r(2)+\cdots+r(n)}{n+1}$, the average value, is exactly equal to $\frac{1}{n+1}$ times the number of lattice points in the quarter disc $x, y \geq 0, x^2+y^2 \leq n$. Consideration of the double Riemann integral show that this average approaches the area of the unit quarter circle, namely $\pi/4$, and so for this set A, $\frac{r(0)+r(1)+r(2)+\cdots+r(n)}{n+1} \to \frac{\pi}{4}$ ($r(n)$ is *on average* equal to the constant $\pi/4$.)

The difficult question is how quickly this limit is approached. Thus fairly simple reasoning shows that

$$\frac{r(0)+r(1)+r(2)+\cdots+r(n)}{n+1} = \frac{\pi}{4} + O\left(\frac{1}{\sqrt{n}}\right),$$

whereas more involved analysis shows that

$$\frac{r(0)+r(1)+r(2)+\cdots+r(n)}{n+1} = \frac{\pi}{4} + O\left(\frac{1}{n^{2/3}}\right).$$

Very deep arguments have even improved this to $o\left(\frac{1}{n^{2/3}}\right)$, for example, and the conjecture is that it is actually $O\left(\frac{1}{n^{\frac{3}{4}-\epsilon}}\right)$ for every $\epsilon > 0$. On the other hand, further difficult arguments show that it is not $O\left(\frac{1}{n^{\frac{3}{4}+\epsilon}}\right)$.

Now all of these arguments were made for the very special case of $A =$ the perfect squares. What a surprise then, when Erdös and Fuchs showed, by *simple* analytic number theory, the following:

Theorem. *For any set* A, $\frac{r(0)+r(1)+r(2)+\cdots+r(n)}{n+1} = C + O\left(\frac{1}{n^{\frac{3}{4}+\epsilon}}\right)$ *is impossible unless* $C = 0$.

This will be proved in the current chapter, but first an appetizer. We prove that $r_-(n)$ can't eventually be constant.

So let us assume that

$$A^2(z) - A(z^2) = P(z) + \frac{C}{1-z}, \tag{1}$$

P is a polynomial, and C is a positive constant. Now look for a contradiction. The simple device of letting $z \to (-1)^+$ which worked so nicely for the r_+ problem, leads nowhere here. The exercised in Chapter I were, after all, hand picked for their simplicity and involved only the *lightest touch* of analysis. Here we encounter a slightly heavier dose. We proceed, namely, by integrating the modulus around a circle. From (1), we obtain, for $0 \leq r < 1$,

$$\begin{aligned} &\int_{-\pi}^{\pi} |a^2(re^{i\theta})|d\theta \\ &\quad \leq \int_{-\pi}^{\pi} |a(r^2e^{2i\theta})|d\theta + \int_{-\pi}^{\pi} |P^2(re^{i\theta})|d\theta \\ &\quad + C\int_{-\pi}^{\pi} \frac{d\theta}{|1-re^{i\theta}|}. \end{aligned} \tag{2}$$

Certain estimates are fairly evident. $P(z)$ is a polynomial and so

$$\int_{-\pi}^{\pi} |P(re^{i\theta})| d\theta \leq M, \tag{3}$$

independent of r.

We can also estimate the (elliptic) integral $\int_{-\pi}^{\pi} \frac{d\theta}{|1-re^{i\theta}|} = 2\int_{0}^{\pi} \frac{d\theta}{|1-re^{i\theta}|}$ by the observation that if z is any complex number in the first quadrant, then $|z| \leq \Re z + \Im z$. Thus since for $0 \leq \theta \leq \pi$, $1 - re^{i\theta} is$ in the first quadrant, $\frac{ie^{i\theta}}{e^{i\theta}-r} = \frac{i}{1-re^{-i\theta}}$ also is, and $\frac{1}{|1-re^{-i\theta}|} = \left|\frac{ie^{i\theta}}{e^{i\theta}-r}\right| \leq (\Re + \Im)\left(\frac{ie^{i\theta}}{e^{i\theta}-r}\right)$. Hence

$$\begin{aligned}\int_{0}^{\pi} \frac{d\theta}{|1-re^{i\theta}|} &\leq (\Re + \Im) \int_{0}^{\pi} \frac{ie^{i\theta}}{e^{i\theta}-r} d\theta \\ &= (\Re + \Im)(\log(e^{i\theta} - r)) \mid |_{0}^{\pi} \\ &= (\Re + \Im) \log\left(-\frac{1+r}{1-r}\right) \\ &= \pi + \log\left(\frac{1+r}{1-r}\right).\end{aligned}$$

The bound, then, is

$$\int_{0}^{\pi} \frac{d\theta}{|1-re^{i\theta}|} \leq 2\pi + 2\log\left(\frac{1+r}{1-r}\right). \tag{4}$$

The integral $\int_{-\pi}^{\pi} |A(re^{i\theta})|^2 d\theta$ is a delight. It succumbs to Parseval's identity. This is the observation that

$$\begin{aligned}\int_{-\pi}^{\pi} |\sum a_n e^{in\theta}|^2 d\theta &= \int_{-\pi}^{\pi} \sum a_n e^{in\theta} \sum \bar{a}_m e^{-im\theta} d\theta \\ &= \int_{-\pi}^{\pi} \sum_{m,n} a_n \bar{a}_m e^{i(n-m)\theta} d\theta \\ &= \sum_{n,m} a_n \bar{a}_m \int_{-\pi}^{\pi} e^{i(n-m)\theta} d\theta\end{aligned}$$

and these integrals all vanish except that, when $n = m$, they are equal to 2π. Hence this double sum is $2\pi \sum |a_n|^2$. The derivation is clearly valid for finite or absolutely convergent series which covers our case of $A(re^{i\theta})$ (but it even holds in much greater "miraculous" generalities).

At any rate, Parseval's identity gives us

$$\int_{-\pi}^{\pi} |A(re^{i\theta})|^2 d\theta = 2\pi \sum_{a \in A} r^{2a} = 2\pi A(r^2). \tag{5}$$

The last integral we must cope with is $\int_{-\pi}^{\pi} |A(r^2 e^{2i\theta})| d\theta$, and, unlike integrals of $|f|^2$, there is no formula for integrals of $|f|$. *But* there is always the Schwarz inequality $\int |f| \le \sqrt{\int 1 \cdot \int |f|^2}$, and so at least we can get an upper bound for such integrals, again by Parseval. The conclusion is that

$$\int_{-\pi}^{\pi} |A(r^2 e^{2i\theta})| d\theta \le 2\pi \sqrt{A(r^4)}. \tag{6}$$

All four of the integrals in (2) have been spoken for and so, by (2) through (6), we obtain

$$A(r^2) \le \sqrt{A(r^4)} + \frac{M}{2\pi} + 1 + \frac{C}{\pi} \log\left(\frac{1+r}{1-r}\right). \tag{7}$$

It is a nuisance that our function A is evaluated at two different points, but we can alleviate that by the obvious monotonicity of A ($A(r^4) \le A(r^2)$) and obtain

$$A(r^2) \le \sqrt{A(r^2)} + M' + \frac{C}{\pi} \log\left(\frac{1+r}{1-r}\right). \tag{8}$$

Is something bounded in terms of its own square root? But if $x \le \sqrt{x} + a$, we obtain $(\sqrt{x} - \frac{1}{2})^2 \le a + \frac{1}{4}$, $\sqrt{x} \le \sqrt{a + \frac{1}{4}} + \frac{1}{2}$, $x \le a + \frac{1}{2} + \sqrt{a + \frac{1}{4}}$. This yields a *pure* bound on x. Then

$$A(r^2) \le M'' + \frac{C}{\pi} \log\left(\frac{1+r}{1-r}\right) + \sqrt{M'' + \frac{C}{\pi} \log\left(\frac{1+r}{1-r}\right)}. \tag{9}$$

But, so what? This says that $A(r^2)$ grows only at the order of $\log \frac{1}{1-r}$ as $r \to 1^-$, but it doesn't say that $A(r^2)$ remains bounded, does it? Wherein is the hoped for contradiction? We must revisit (1) for this. Thereby we obtain, in turn $A^2(r^2) - A(r^4) = P(r^2) +$

$\frac{C}{1-r^2}$, $A^2(r^2) \geq P(r^2) + \frac{C}{1-r^2}$, $A^2(r^2) \geq -M + \frac{C}{1-r^2}$, and finally

$$A(r^2) \geq \sqrt{-M + \frac{C}{1 - r^2}}, \qquad (10)$$

a rate of growth which flatly contradicts (9) and so gives our desired contradiction.

If this proof seems like just so much sleight of hand, let us observe what is "really" going on. We find ourselves with a set A whose $r_i(n)$ is "almost" constant and this means that $A^2(z) \approx \frac{C}{1-z}$. On the one hand, this forces $A(z)$ to be large on the positive axis $\left(A(r^2) > \frac{C'}{\sqrt{1-r^2}}\right)$, and, on the other hand Parseval says that the integral of $|A^2(z)|$ *is* $A(r^2)$ and $\left|\frac{C}{1-z}\right|$ (being fairly small except near 1) has a small integral, only $O(\log \frac{1}{1-r})$. (So $A(r^2) < C'' \log \frac{1}{1-r}$).

In cruder terms, Parseval tells us that $A^2(z)$ is large on average, so it must be large elsewhere than just near $z = 1$, and so it cannot really be like $\frac{C}{1-z}$. (Note that the "elsewhere" in the earlier $r_+(n)$ problem was the locale of -1, and so even that argument seems to be in this spirit.)

So let us turn to the Erdös–Fuchs theorem with the same strategy in mind, viz., to bound $A(r^2)$ below by $\frac{C'}{\sqrt{1-r^2}}$ for obvious reasons and then to bound it above by Parseval considerations.

Erdös–Fuchs Theorem

We assume the A is a set for which

$$r(0) + r(1) + \cdots + r(n) = C(n + 1) + O(n^\alpha), \quad C > 0, \ (11)$$

and we wish to deduce that $\alpha \geq \frac{1}{4}$. As usual, we introduce the generating function $A(z) = \sum_{a\in A} z^a$, so that $A^2(z) = \sum r(n)z^n$, and therefore $\frac{1}{1-z}A^2(z) = \sum[r(0) + r(1) + \cdots + r(n)]z^n$. Since $\sum(n + 1)z^n = \frac{1}{(1-z)^2}$ our hypothesis (11) can be written as

$$\frac{1}{1 - z}A^2(z) = \frac{C}{(1 - z)^2} + \sum a_n z^n, \quad A_n = O(n^\alpha),$$

or

$$A^2(z) = \frac{C}{1-z} + (1-z)\sum a_n z^n, \quad A_n = O(n^\alpha). \tag{12}$$

Of course we may assume throughout that $\alpha < 1$. Thereby (12) yields the bound $M(1-r^2)^{-\alpha-1}$ for $\sum a_n r^{2n}$, so that we easily achieve our first goal namely,

$$A(r^2) > \frac{C'}{\sqrt{1-r^2}}, \quad C' > 0. \tag{13}$$

As for the other goal, the Parseval upper bound on $A(r^2)$, again we wish to exploit the fact that $A^2(z)$ is "near" $\frac{C}{1-z}$, but this takes some doing. From the look of (12) unlike (1), this "nearness" seems to occur only where $(1-z)\sum A_n z^n$ is relatively small, that is, only in a neighborhood of $z = 1$. We must "enhance" this locale if we are to expect anything from the integration, and we do so by multiplying by a function whose "heft" or largeness is all near $z = 1$. A handy such multiplier for us is the function $S^2(z)$ where

$$S(z) = 1 + z + z^2 + \cdots + z^{N-1}, \quad N \text{ large}. \tag{14}$$

The multiplication of $S^2(z)$ by (12) yields

$$[S(z)A(z)]^2 = \frac{CS^2(z)}{1-z} + (1-z^N)S(z)\sum a_n z^n, \tag{15}$$

which gives

$$|S(z)A(z)|^2 \le \frac{CN^2}{|1-z|} + 2|S(z)\sum a_n z^n|, \tag{16}$$

and integration leads to

$$\begin{aligned} \int_{-\pi}^{\pi} |S(re^{i\theta})A(re^{i\theta})|^2 d\theta & \\ \le CN^2 \int_{-\pi}^{\pi} \frac{d\theta}{|1-re^{i\theta}|} & \qquad (17) \\ + 2\int_{-\pi}^{\pi} |S(re^{i\theta})\sum a_n (re^{i\theta})^n| d\theta. & \end{aligned}$$

As before, we will use Parseval on the first of these integrals, (4) on the second, and Schwarz's inequality together with Parseval on the third.

So write $S(z)A(z) = \sum C_n z^n$, and conclude that $\int_{-\pi}^{\pi} |S(re^{i\theta}) A(re^{i\theta})|^2 d\theta = 2\pi \sum |c_n|^2 r^{2n}$. Since the c_n are integers, $|c_n|^2 = c_n^2 \geq c_n$ and so this is, furthermore, $\geq 2\pi \sum c_n r^{2n} = S(r^2)A(r^2)$. (The general fact then is that, if $F(z)$ has integral coefficients, $\int_{-\pi}^{\pi} |F(re^{i\theta})|^2 d\theta \geq 2\pi F(r^2)$.)

Now we introduce a side condition on our parameters r and N which we shall insist on henceforth namely that

$$\frac{1}{1-r^2} \geq N. \tag{18}$$

Thus, by (14), $S(r^2) > Nr^{2N} \geq N(1-\frac{1}{N})^N \geq N(1-\frac{1}{2})^2 = \frac{N}{4}$, and by (13), $A(r^2) > \frac{C'}{\sqrt{1-r^2}}$, and we conclude that

$$\int_{-\pi}^{\pi} |S(re^{i\theta})A(re^{i\theta})|^2 d\theta > \frac{C''N}{\sqrt{1-r^2}}, \qquad C'' > 0. \tag{19}$$

Next, (4) gives

$$CN^2 \int_{-\pi}^{\pi} \frac{d\theta}{|1-re^{i\theta}|} \leq MN^2 \log \frac{e}{1-r^2} \tag{20}$$

and our last integral satisfies

$$\begin{aligned}
&\int_{-\pi}^{\pi} |S(re^{i\theta}) \sum a_n (re^{i\theta})^n| d\theta \\
&\leq \sqrt{\int_{-\pi}^{\pi} |S(re^{i\theta})|^2 d\theta \int_{-\pi}^{\pi} |\sum a_n (re^{i\theta})^n|^2 d\theta} \\
&= 2\pi \sqrt{\sum_{k<N} r^{2k} \sum |a_n|^2 r^{2n}} \leq 2\pi \sqrt{N} M \sqrt{\sum n^{2\alpha} r^{2n}}.
\end{aligned}$$

Applying (13) and (14) again leads finally to

$$\int_{-\pi}^{\pi} |S(re^{i\theta}) \sum a_n (re^{i\theta})^n| d\theta \leq \frac{M\sqrt{N}}{(1-r^2)^{\alpha+\frac{1}{2}}} \tag{21}$$

At last, combining (19), (20), and (21) allows the conclusion

$$\frac{C''}{M} \leq N\sqrt{1-r^2}\log\frac{e}{1-r^2} + \frac{1}{\sqrt{N}(1-r^2)^{\alpha}}. \tag{22}$$

Once again we are masters of the parameters (subject to (18)), and so we elect to choose r, so that $N\sqrt{1-r^2} = \frac{1}{\sqrt{N}(1-r^2)^{\alpha}}$. Thus our choice is to make $\frac{1}{1-r^2} = N^{\frac{3}{2\alpha+1}}$ and note happily that our side condition (18) is satisfied. Also "plugging" this choice into (22) gives

$$\frac{C''}{M} \leq N^{\frac{4\alpha-1}{4\alpha+2}}(2+3\log N). \tag{23}$$

Well, success is delicious. We certainly see in (23) the fact that $\alpha \geq \frac{1}{4}$. (If the exponent of N, $\frac{4\alpha-1}{4\alpha+2}$, were negative then this right-hand side would go to 0, $2+3\log N$ notwithstanding, and (23) would become false for large N.)

Problems for Chapter III

1. Show that the number of lattice points in $x^2 + y^2 \leq n^2, x, y \geq 0$, is $\sim \frac{\pi}{4} n^2$. By the Riemann integral method show that it is, in fact $= \frac{\pi}{4} n^2 + O(n)$.

2. If x is bounded by its own square root (i.e., by $\sqrt{x} + a$), then we find that it has a pure bound. What if x, instead, is bounded by $x^{2/3} + ax^{1/3} + b$? Does this insure a bound on x?

3. Suppose that a convex closed curve has its curvature bounded by δ. Show that it must come within $2\sqrt{\delta}$ of some lattice point.

4. Produce a convex closed curve with curvature bounded by δ which doesn't come within $\frac{\sqrt{\delta}}{1200}$ of any lattice point.

IV

Sequences without Arithmetic Progressions

The gist of the result of Chapter IV is that a sequence of integers with "positive density" must contain an arithmetic progression (of at least three distinct terms).

More precisely and in sharper, finitized form, this is the statement that, if $\epsilon > 0$, then for large enough n, any subset of the nonnegative integers below n with at least ϵn members must contain three terms a, b, c where $a < b < c$ and $a + c = 2b$. This is a shock to nobody. If a set is "fat" enough, it *should* contain all sorts of patterns. The shock is that this is so hard to prove.

At any rate we begin with a vastly more general consideration, the notion of an "affine property" of finite sets of integers. So let us agree to call a property P an *affine property* if it satisfies the following two conditions:

1. For each fixed pair of integers α, β with $\alpha \neq 0$, the set a_n has P if and only if $\alpha a_n + \beta$ has P.
2. Any subset of a set, which has P, also has P.

Thus, for example, the property P_A of not containing any arithmetic progressions is an affine property. Again the trivial property P_0 of just being *any* set is an affine one.

Now we fix an affine property P and consider a *largest* subset of the nonnegative integers below n, which has P. (Thus we require that this set has the most members possible, not just to be maximal.) There may be several such sets but we choose one of them and denote it by $S(n; P)$. We also denote the number of elements of this set by

$f(n;\, P)$. So, for example, for the trivial property, $f(n;\, P_0) = n$, and for P_A, $f(3;\, P_A) = 2$, $f(5;\, P_A) = 4$.

It follows easily from conditions 1 and 2 that this $f(n)$ is subadditive, i.e., $f(m+n) \leq f(m) + f(n)$. If we recall the fact that subadditive functions enjoy the property that $\lim_{n\to\infty} \frac{f(n)}{n}$ exists (in fact $\lim_{n\to\infty} \frac{f(n)}{n} = \inf \frac{f(n)}{n}$), we are led to define $C_p = \lim_{n\to\infty} \frac{f(n;P)}{n}$. This number is a measure of how *permissive* the property P is. Thus $C_{P_0} = 1$, because P_0 is *totally* permissive. The announced result about progression = free sequences amounts to the statement that $C_{P_A} = 0$, so that P_A is, in this sense, totally *unpermissive*. At any rate, we always have $0 \leq C_P \leq 1$, and we may dub C_P the *permission* constant.

The remarkable result proved by Szemeredi and then later by Furstenberg is that, except for P_0, C_P is *always* 0. Their proofs are both rather complicated, and we shall content ourselves with the case of P_A, which was proved by Roth.

The Basic Approximation Lemma

It turns out that the extremal sets $S(n;\, P)$ all behave very much as though their elements were chosen at random. For example, we note that such a set must contain roughly the same number of evens as odds. Indeed if $2b_1, 2b_2, \cdots, 2b_k$ were its even elements, then $b_1, b_2, \cdots, b_k$ would be a subset of $(0, \frac{n}{2})$ and so we could conclude that $k \leq f(\frac{n}{2})$. Similarly the population of the odd elements of S would satisfy this same inequality. Since $\frac{n}{2} \sim \frac{1}{2} f(n)$, we conclude that both the evens and the odds contain *not much more* than half the whole set. Thereby the evens and the odds must be roughly equinumerous. (Thus, *two* upper bounds imply the lower bounds.)

Delaying for the moment the precise statement of this "randomness," let us just note how it will prove useful to us with regard to our arithmetic progression considerations. The point is simply that, if integers were chosen truly at random with a probability $C > 0$, there would automatically be a huge number of arithmetic progressions formed. So we expect that even an approximate randomness should produce at least one arithmetic progression.

The precise assertion is that of the following lemma.

Lemma. $\sum_{a\in S(n;P)} z^a = C_p \sum_{k\leq n} z^k + o(n)$, *uniformly on* $|z| = 1$.

Remark. In terms of the great Szemeredi–Furstenberg result that $C_P \equiv 0$ (except for $P = P_0$), this is a total triviality. We are proving what in truth is an empty result. Nevertheless we are not prepared to give the lengthy and complex proofs of this general theorem, and so we must prove the Lemma. (We do what we can.) The proof, in fact, is really just an elaboration of the odds and evens considerations above.

PROOF. The basic strategy is to estimate $q(z) = \sum_{a\in S} z^a - C_P \sum_{k<n} z^k$, *together with all* of its partial sum at every root of unity of order up to N (N is a parameter to be chosen later). The point is that, if we have a bound on a polynomial *and its partial sums* at a point, then we inherit a bound on that polynomial throughout an arc around that point. (Thereby we will obtain bounds for arcs between the roots of unity which will fill up the whole circle.)

Specifically, we have the identity

$$\frac{p(z)}{1-\frac{z}{\zeta}} = \sum_{m<n} p_m(\zeta)\left(\frac{z}{\zeta}\right)^m + \frac{p(z)}{1-\frac{z}{\zeta}}\left(\frac{z}{\zeta}\right)^n, \tag{1}$$

for any polynomial p of degree at most n, where the p_m denote the partial sums. (This simply records the result of the "long division.")

From (1) we easily obtain the bound $|p(z)| \leq |\zeta - z| \sum_{m<n} |p_m(\zeta)| + |p(\zeta)|$, and so we conclude the following:

If all the partial sums are bounded by M at ζ, the polynomial is bounded by $M(n\ell + 1)$ throughout an arc of length 2ℓ centered at ζ. (2)

So let $\alpha \leq N$ be chosen, and let ω be any αth root of unity, i.e., $\omega^\alpha = 1$. To estimate $q_m(\omega)$, let us write it as

$$\sum_{\beta=1}^{\alpha} \omega^\beta \left(\sum_{\substack{a\in S\\ a<m\\ a\equiv\beta(\alpha)}} 1 - C_P \sum_{\substack{k<m\\ k\equiv\beta(\alpha)}} 1 \right),$$

and let us note that the first inner sum

$$\sigma_\beta = \sum_{\substack{a\in S\\ a<m\\ a\equiv\beta(\alpha)}} 1$$

counts the seze of a subset of S, which therefore has P which is affine to a subset of $(0, \frac{m}{\alpha})$, and so has at most $f(\frac{m}{\alpha})$ elements (where we write $f(x)$ for $f(\lceil x \rceil)$).

Thus

$$\begin{aligned}
q_m(\omega) &= -\sum_{\beta=1}^{\alpha} \omega^\beta \left(f\left(\frac{m}{\alpha}\right) - \sigma_\beta \right) \\
&\quad + \sum_{\beta=1}^{\alpha} \omega^\beta \left[f\left(\frac{m}{\alpha}\right) - C_P \sum_{\substack{k<m\\ k\equiv\beta(\alpha)}} 1 \right] \\
&\leq \sum_{\beta=1}^{\alpha} \left| f\left(\frac{m}{\alpha}\right) - \sigma_\beta \right| + \sum_{\beta=1}^{\alpha} \left| f\left(\frac{m}{\alpha}\right) - C_P \frac{m}{\alpha} \right| \qquad (3) \\
&= \sum_{\beta=1}^{\alpha} \left(f\left(\frac{m}{\alpha}\right) - \sigma_\beta \right) + \sum_{\beta=1}^{\alpha} \left(f\left(\frac{m}{\alpha}\right) - C_P \frac{m}{\alpha} \right) \\
&= 2\alpha f\left(\frac{m}{\alpha}\right) - \sum_{\beta=1}^{\alpha} \sigma_\beta - C_P m.
\end{aligned}$$

If we next note that $\sum_{\beta=1}^{\alpha} \sigma_\beta$ is exactly the number of elements of S which are below m and so is equal to $f(n)$ minus the number of elements of S which are $\geq m$, we obtain

$$\sum_{\beta=1}^{\alpha} \sigma_\beta \geq f(n) - f(n-m) \geq C_P n - f(n-m). \qquad (4)$$

Substituting (4) in (3) gives

$$q_m(\omega) \ll 2\alpha\left[f\left(\frac{m}{\alpha}\right) - C_P\frac{m}{\alpha}\right] + (f(n-m) - C_P(n-m)). \quad (5)$$

Now we find it useful to replace the function $f(x) - C_P x$ by its "monotone majorant" $F(x) - \max_{t\le x}(f(t) - C_P t)$ and note that this $F(x)$ is nondecreasing and satisfies $F(x) = o(x)$ since $f(x) - C_P x$ satisfies the same. So (5) can be replaced by

$$q_m(\omega) \ll 2\alpha F\left(\frac{m}{\alpha}\right) + F(n-m) \le 2\alpha F\left(\frac{n}{\alpha}\right) + F(n) \quad (6)$$

(a bound dependent of m).

So choose n_0 so that $s \ge x_0$ implies $F(x) \le \epsilon x$, and then choose n_1 so that $x \ge n_1$ implies $F(x) \le \frac{\epsilon}{n_0}x$. From now on we will pick $n \ge N - 1$ and also will fix $N = [\frac{n}{n_0}]$.

Dirichlet's theorem[1] on approximation by rationals now tells us that the totality of arcs surrounding these ω with length $2\frac{2\pi}{\alpha(N+1)}$ covers the whole circle. Thus using (2) for $q(z)$, $\zeta = \omega$ and $\ell = 2\frac{2\pi}{\alpha(N+1)} \le \frac{2\pi n_0}{n_\alpha}$ gives

$$q(z) \ll [2\alpha F\left(\frac{n}{\alpha}\right) + F(n)]\left(1 + 2\pi\frac{n_0}{\alpha}\right). \quad (7)$$

We separate two cases:

Case I: $\alpha \le n_0$. Here we use $F(\frac{n}{\alpha}) \le F(n)$ and obtain $[2\alpha F(\frac{n}{\alpha}) + F(n)](1+\frac{2\pi n_0}{\alpha}) \le (2\alpha+1)(1+\frac{2\pi n_0}{\alpha})F(n) \le 3\alpha(1+\frac{2\pi n_0}{\alpha})F(n) = (3\alpha + 6\pi n_0)F(n) \le (6\pi + 3)n_0 F(n) \le (6\pi + 3)\frac{\epsilon}{n_0}n \le 22\epsilon n$.

Case II: $\alpha > n_0$. Here $[2\alpha F(\frac{n}{\alpha}) + F(n)](1 + \frac{2\pi n_0}{\alpha}) \le [2\alpha F(\frac{n}{\alpha}) + F(n)](1 + 2\pi)$. But still $\alpha \le \frac{n}{n_0}$, or $\frac{n}{\alpha} \ge n_0$. So $F(\frac{n}{\alpha}) \le \epsilon\frac{n}{\alpha}$, and the above is $\le (2\epsilon n + \epsilon n)(1 + 2\pi) = (3 + 6\pi)\epsilon n < 22\epsilon n$.

In either case Dirichlet's theorem yields our lemma.

So let P be *any* affine property, and denote by $\mathcal{A} = \mathcal{A}(n;\, P)$ the number of arithmetic progressions from $S(n;\, P)$(where order counts

---

[1]Dirichlet's theorem can be proved by considering the powers $1, z, z^2, \cdots, z^N$ for z any point on the unit circle. Since these are $N + 1$ points on the circle, two of them z^i, j^j must be within arc length $\frac{2\pi}{N+1}$ of one another. This means $|\arg Z^{i-j}| \le \frac{2\pi}{N+1}$ and calling $|i - k| = \alpha$ gives the result.

and equality is allowed). We show that

$$\mathcal{A}(n;\, P) = \frac{C_P^3}{2} n^2 + o(n^2). \tag{8}$$

The proof is by contour integration. If we abbreviate $\sum_{a \in S} z^a = g(z)$, then we recognize $\mathcal{A}$ as the constant term in $g(z)g(z)g(z^{-2})$, and so we may write

$$\mathcal{A} = \frac{1}{2\pi i} \int_{|z|=1} g^2(z) g(z^{-2}) \frac{dz}{z}. \tag{9}$$

Now writing $G(z) = \sum_{k<n} z^k$, $g(z) = C_P G(z) + q(z)$ (where q is "small" by the lemma). If we substitute this in (9), we obtain

$$C_P^3 \frac{1}{2\pi i} \int_{|z|=1} G^2(z) G(z^{-2}) \frac{dz}{z}$$

plus seven other integrals. Each of these *other* integrals is the product of three functions, each a G or a q, and at least one of them is a q. By our lemma, then, we may estimate each of these seven integrals by $o(n)$ times an integral of the product of *two* functions. Both of these functions are either a $|G|$ or a $|q|$. As such each is estimable by the Schwarz inequality, Parseval equality techniques. The final estimate for each of these seven integrals, therefore, is $o(n)\sqrt{nn} = 0(n^2)$, and so (9) gives

$$\mathcal{A} = C_P^3 \frac{1}{2\pi i} \int_{|z|=1} G^2(z) G(z^{-2}) \frac{dz}{z} + o(n^2). \tag{10}$$

But reading (9) for the property P_0 shows that this integral is simply $\mathcal{A}(n;\, P_0)$ and it is a simple exercise to show that $\mathcal{A}(n;\, P_0)$, the number of triples below n which are in arithmetic progression, is exactly $\lceil \frac{n^2}{2} \rceil$. Indeed, then (10) reduces to (8).

All of our discussion thus far has been quite general and is valid for arbitrary affine properties. We finally become specific by letting $P = P_{\mathcal{A}}$, and we easily deduce the following:

Theorem (Roth). $C_{P_{\mathcal{A}}} = 0$.

PROOF. By the definition of $P_{\mathcal{A}}$, the only arithmetic progressions in $S(n;\ P_{\mathcal{A}})$ are the trivial ones, three equal terms, which number at most n. Thus $\mathcal{A}(n;\ P_{\mathcal{A}}) \leq n$, and so, by (8), $C_{P_{\mathcal{A}}}^3 \frac{n^2}{2} + o(n^2) \leq n$. Therefore $C_{P_{\mathcal{A}}} = 0$. Q.E.D.

Problems for Chapter IV

1. Attach a positive rational to each integer from 1 to 12 so that all A.P.'s with common difference d up to 6 obtain their "correct" measure $\frac{1}{d}$.

2. Prove that, if we ask for a generalization of this, then we can only force the correct measdure $\frac{1}{d}$ for *all* A.P.'s of common difference d, by attaching weights onto $1, 2, \ldots, n$, if $d = O(\sqrt{n})$.

3. If we insist only on approximation, however, show that we can always attach weights onto $1, 2, \ldots, n$ such that the "measure" given to every A.P. with common difference $\leq m$ is within $e^{-n/m}$ of $\frac{1}{a}$.

V

The Waring Problem

In a famous letter to Euler, Waring wrote his great conjecture about sums of powers. Lagrange had already proved his magnificent theorem that every positive integer was the sum of four squares.

Waring guessed that this was not just a property of squares, but that, in fact, the sum of a fixed number of cubes, fourth powers, fifth powers, etc., also worked. He guessed that every positive integer was the sum of 9 cubes, 19 fourth powers, 37 fifth powers, and so forth, and although no serious guess was made as to how the sequence 4 (squares), 9, 19, 37, ... went on, he simply stated that it *did*! That is what we propose to do in this chapter, just to prove the existence of the requisite number of the cubes, fourth powers, etc. We do not attempt to find the structure of the 4, 9, 19, ..., but just to prove its existence.

So let us fix k and view the kth powers. Our aim, by Schnirelmann's lemmas below, need be only to produce a $g = g(k)$ and an $\alpha = \alpha(k) > 0$ such that the sum of $g(k)$ kth powers represents at least the fraction $\alpha(k)$ of all of the integers.

One of the wonderful things about this approach is that it requires only upper bounds, despite the fact that Waring's conjecture seems to require lower bounds, something seemingly totally impossible for contour integrals to produce. But the adequate upper bounds are obtained by the so called Weyl sums given below.

So first we turn to our three basic lemmas which will eventually yield our proof. These are A, the theorem of Dirichlet, B, that of Schnirelmann, and finally C, the evaluation of the Weyl sums.

A. Theorem (Dirichlet). *Given a real x and a positive integer M, there exists an integer a and a positive integer $b \leq M$ such that* $|x - \frac{a}{b}| \leq \frac{1}{(M+1)b}$.

PROOF. Consider the numbers $0, x, 2x, 3x, \ldots, Mx$ all reduced (mod 1). Clearly, two of these must be within $\frac{1}{M+1}$ of each other. If these two differ by bx, then $1 \leq b \leq M$ and bx (mod 1) is, in magnitude, $\leq \frac{1}{M+1}$. Next pick an integer a that makes $bx - a$ equal to bx (mod 1). So $|bx - a| \leq \frac{1}{(M+1)}$ which means $|x - \frac{a}{b}| \leq \frac{1}{(M+1)b}$, as asserted.

We also point out that this is a best possible result as the choice $x = \frac{1}{M+1}$ shows for every M. (Again, we may assume that $(a, b) = 1$ for, if they have a common divisior, this would make the inequality $|b| \leq M$ even truer).

B. Schnirelmann's Theorem. *If S has positive density, then there is a fixed integer k, such that every integer is the sum of at most k members of S.*

Lemma 1. *Let S have density α. Then $S \oplus S$ has density at least $2\alpha - \alpha^2$.*

PROOF. All the gaps in the set S are covered in part by the translation of S by the term of S just before this gap. Hence, at least the fraction α of this gap gets covered. So from this covering we have density α from S itself and α times the gaps. Altogether, then, we indeed have $\alpha + \alpha(1 - \alpha) = 2\alpha - \alpha^2$, as claimed. Q.E.D.

Lemma 2. *If S has density $\alpha > \frac{1}{2}$, then $S \oplus S$ contains all the positive integers.*

PROOF. Fix an integer n which is arbitrary, let A_n be the subset of S which lies $\leq n$, and let B be the set of all n minus elements of S. Since A contains more than $n/2$ elements and B contains at least $n/2$ elements, the Pigeonhole principle guarantees that they overlap. So suppose they overlap at k. Since $k \in A$, we get $k \in S$, and since

$k \in B$, we get $n - k \in S$. These are the two elements of S which sum to n.

Repeating Lemma 1 j times, then, leads to a summing of 2^j copies of S and a density of $1 - (1 - \alpha)^{2^j}$ or more. Since this latter quantity, for large enough j, will become bigger than $\frac{1}{2}$, Lemma 2 tells us that 2^{j+1} copies of S give us all the integers, just as Schnirelmann's theorem claims.

C. Evaluation of Weyl Sums. *Let $k \leq N$, $P(n)$ be a polynomial of degree k with real coefficients and leading coefficient integral and prime to b, and let I be an interval of length $\leq N$. Then*

$$\sum_{n \in I} e\left(\frac{P(n)}{b}\right) \ll N^{1+o(1)} b^{-2^{1-k}}.$$

We proceed by induction on k, which represents the degree of $P(n)$. It is clearly true for $k = 1$. and generally we may write

$$\sum_{n \in I} e\left(\frac{P(n)}{b}\right)$$

and may assume w.l.o.g. that $I = \{1, 2, 3, \ldots, N\}$. Thereby

$$|s|^2 = \sum_{j=-N}^{N} \sum_{\substack{n \in \{1,2,\ldots,N\} \\ n \in \{j+1, j+2, \ldots, j+N\}}} e\left(\frac{P(n) - P(n-j)}{b}\right).$$

This inner sum involves a polynomial of degree $(k - 1)$ but has a leading coefficient which varies with j. If we count those j which produce a denominator of d, which of course must divide b, then we observe that this must appear roughly d times in an interval of length b. So this number of j in the full interval of length $2N + 1$ is roughly $\frac{(2N+1)}{b} d$.

The full estimate, then, by the inductive hypothesis is

$$|S|^2 \leq \sum_{d \mid b} \frac{N}{b} d N^{1+o(1)} d^{-\frac{1}{2^{k-2}}} \leq \frac{N^{2+o(1)}}{b} b^{1-\frac{1}{2^{k-2}}} \sum_{d \mid b} 1$$

$$\leq N^{2+o(1)} b^{-\frac{1}{2^{k-2}}} b^{o(1)}.$$

So we obtain

$$|S| \leq N^{1+o(1)} b^{-\frac{1}{2^{k-1}}},$$

and the induction is complete.

Now we continue as follows:

Lemma 3. *Let $k > 1$ be a fixed integer. There exists a C_1 such that, for any positive integers N, a, b with $(a, b) = 1$,*

$$\left| \sum_{n=1}^{N} e(\frac{a}{b} n^k) \right| \leq C_1 N^{1+o(1)} b^{-2^{1-k}}.$$

(Throughout, we write $e(t) = e^{2\pi i t}$, and C_1, C_2 denote constants.) Our endpoint will be the following:

Theorem. *If, for each positive integer s, we write*

$$r_s(n) = \sum_{\substack{n_1^k + \cdots + n_s^k = n \\ n_i \geq 0}} 1,$$

then there exists g and C such that $r_g(n) \leq C n^{g/k-1}$ for all $n > 0$.

The previously cited notions of Schnirelmann allow deducing, the full Waring result from this theorem:

There exists a G for which $r_G(n) > 0$ for all $n > 0$.

To prove our theorem, since

$$r_g(n) = \int_0^1 \left[\sum_{m \leq n^{1/k}} e(x m^k) \right]^g e(-nx) dx,$$

it suffices to prove that there exists g and C for which

$$\int_0^1 \left| \sum_{n=1}^{N} e(x n^k) \right|^g dx \leq C N^{g-k} \quad \text{for all} \quad n > 0. \tag{1}$$

First some parenthetical remarks about this inequality. Suppose it is known to hold for some C_0 and g_0. Then, since $|\sum_{n=1}^{N} e(x n^k)| \leq N$, it persists for C_0 and any $g \geq g_0$. Thus (1) is a property of large g's, in other words, it is purely a "magnitude property." Again, (1) is a best possible inequality in that, for each g, there exists a $c > 0$ such that

$$\int_0^1 \left|\sum_{n=1}^{N} e(xn^k)\right|^g dx > cN^{g-k} \quad \text{for all} \quad n > 0. \tag{2}$$

To see this, note that $\sum_{n=1}^{N} e(xn^k)$ has a derivative bounded by $2\pi N^{k+1}$. Hence, in the interval $(0, \frac{1}{4\pi N^k})$,

$$\left|\sum_{n=1}^{N} e(xn^k)\right| \geq N - 2\pi N^{k+1}\frac{1}{4\pi N^k} = \frac{N}{2},$$

and so (2) follows with $c = \frac{1}{4\pi 2^g}$.

The remainder of our paper, then, will be devoted to the derivation of (1) from Lemma 3. Henceforth k is fixed. Denote by $I_{a,b,N}$ the x-interval $|x - \frac{a}{b}| \leq \frac{1}{bN^{k-1}}$, and call $J = N^k|x - \frac{a}{b}|$, $j = [J]$, where a, b, N, j are integers satisfying $N > 0, b > 0, 0 \leq a < b$, $(a, b) = 1, b \leq N^{k-\frac{1}{2}}$.

By Dirichlet's theorem, these intervals cover $(0, 1)$. Our main tool is the following lemma:

Lemma 4. *There exists $\epsilon > 0$ and C_2 such that, throughout any interval $I_{a,b,N}$,*

$$\left|\sum_{n=1}^{N} e(xn^k)\right| \leq \frac{C_2 N}{(b + j)^\epsilon}.$$

PROOF. This is almost trivial if $b > N^{2/3}$, for, since the derivative of $|\sum_{n=1}^{N} e(xn^k)|$ is bounded by $2\pi N^{k+1}$,

$$\left|\sum_{n=1}^{N} e(xn^k)\right| \leq \left|\sum_{n=1}^{N} e(\frac{a}{b}n^k)\right| + \left|x - \frac{a}{b}\right| 2\pi N^{k+1}$$
$$\leq \frac{N^{1+o(1)}}{b^{\frac{1}{2^{k-1}}}} + \frac{2\pi N^{3/2}}{b} (\text{by C}) \leq \frac{N^{1+o(1)}}{b^{\frac{1}{2^{k-1}}}} + \frac{2\pi N}{b^{1/4}},$$

which gives the result, since $j = 0$ automatically. Assume therefore that $b \leq N^{2/3}$, and note the following two simple facts (for details see [K. Knopp, *Theory and Application of Infinite Series*, Blackie & Sons, Glasgow, 1946.] and [G. Pólya und G. Szegö, *Aufgaben und Lehrsátze aus der Analysis*, Dover Publications, New York 1945, Vol. 1, Part II, p. 37]). Q.E.D.

Next, we observe the following two simple exercises, (A) and (B). (A) If M is the maximum of the moduli of the partial sums $\sum_{n=1}^{m} a_n$, V the total variation of $f(t)$ in $0 \le t \le N$, and M' the maximum of the modulus of $f(t)$ in $0 \le t \le N$, then

$$\left|\sum_{n=1}^{N} a_n f(n)\right| \le M(V + M').$$

(B) If V is the total variation of $f(t)$ in $0 \le t \le N$, then

$$\left|\sum_{n=1}^{N} f(n) - \int_0^N f(t)dt\right| \le V.$$

Now write $\alpha = \frac{1}{b}\sum_{n=1}^{b} e(\frac{a}{b}n^k)$ and

$$\sum_{n=1}^{N} e(xn^k) = S_1 + \alpha S_2, \tag{3}$$

where

$$S_1 = \sum_{n=1}^{N}\left[e\left(\frac{a}{b}n^k\right) - \alpha\right] e\left[\left(x - \frac{a}{b}\right)n^k\right],$$

$$S_2 = \sum_{n=1}^{N} e\left[\left(x - \frac{a}{b}\right)n^k\right].$$

We apply (A) to S_1. To do so, we note that

$$\left|\sum_{n=1}^{m}\left[e\left(\frac{a}{b}n^k\right) - \alpha\right]\right| = \left|0 + \sum_{b[m/b]<n\le m}\left[e\left(\frac{a}{b}n^k\right) - \alpha\right]\right|$$
$$\le (1 + |\alpha|)b \le 2b.$$

Also, the total variation of $e[(x-\frac{a}{b})t^k]$ is equal to $2\pi|x-\frac{a}{b}|N^k \leq \frac{2\pi\sqrt{N}}{b}$, whereas $M'=1$. The result is

$$|S_1| \leq 4\pi\sqrt{N}+2b \leq 5\pi n^{2/3}. \tag{4}$$

Next we apply (B) to S_2 and obtain

$$|S_2| \leq \left|\int_0^N e\left[\left(x-\frac{a}{b}\right)t^k\right]dt\right| + \frac{2\pi\sqrt{N}}{b}. \tag{5}$$

Since

$$\left|\int_0^N e\left[\left(x-\frac{a}{b}\right)t^k dt\right]\right| = \left|\int_0^N e\left(\left(x-\frac{a}{b}\right)t^k\right)dt\right|$$
$$= \frac{N}{J^{1/k}}\int_0^{J^{1/k}} e(u^k)du \leq \frac{NC_3}{J^{1/k}},$$

$\int_0^\infty e(u^k)du$ converges.

Combining this with (5) gives

$$|\alpha S_2| \leq \frac{C_4 N|\alpha|}{(1+j)^{1/k}} + 2\pi\sqrt{N}. \tag{6}$$

Now if we apply C to the case $N=b$, we obtain $|\alpha| \leq \frac{C_1}{b^\delta}$, $\delta = 2^{1-k}$, and by (3) the addition of (4) and (6) gives

$$\left|\sum_{n=1}^N e(xn^k)\right| \leq \frac{C_5 N}{b\delta(1+j)^{1/k}} + 7\pi N^{2/3}$$
$$\leq \frac{C_5 N}{b\delta(1+j)^{1/k}} + \frac{C_6 N}{(b+j)^{1/2}}.$$

Since $j \leq \sqrt{N}$ and $b \leq N^{2/3}$, the choice $C+2 = C_5+C_6+C_1+2\pi$, $\epsilon = \min(\frac{\delta}{k}, \frac{1}{k}, \frac{1}{4})$ completes the proof.

Proof of (1). Choose $g \geq \frac{4}{\epsilon}$, ϵ given as above. By Lemma 4, since the length of each $I_{a,b,N}$ is at most $2N^{-k}$,

$$\int_{I_{a,b,N}} \left|\sum_{n=1}^N e(xn^k)\right|^g dx \leq \frac{C_7 N^g}{(b+j)^4}\frac{1}{N^k}.$$

Summing over all a, b, j gives the estimate

$$C_7 N^{g-k} \sum_{b,j} \frac{b}{(b+j)^4} \leq C N^{g-k}.$$

since $\sum_{b=1}^{\infty} \sum_{j=0}^{\infty} \frac{1}{(b+j)^3} < \infty$, and the proof is complete.

Problems for Chapter V

1. If we permit polynomials with arbitrary complex coefficients and ask the "Waring" problem for polynomials, then show that x is *not* the sum of 2 cubes, but it is the sum of 3 cubes.

2. Show that *every* polynomial is the sum of 3 cubes.

3. Show, in general, that the polynomial x is "pivotal," that if x is the sum of g nth powers, then every polynomial is the sum of g nth powers.

4. Show that if $\max(z, b) > 2c$, where c is the degree of $R(x)$, then $P^a + Q^b = R$ is unsolvable.

5. Show that the constant polynomial 1 can be written as the sum of $\sqrt{4n+1}$ nth powers of nonconstant polynomials.

VI

A "Natural" Proof of the Nonvanishing of L-Series

Rather than the usual adjectives of "elementary" (meaning not involving complex variables) or "simple" (meaning not having too many steps) which refer to proofs, we introduce a new one, "natural." This term, which is just as undefinable as the others, is introduced to mean not having any ad hoc constructions or *brilliancies*. A "natural" proof, then, is one which proves itself, one available to the "common mathematician in the streets."

A perfect example of such a proof and one central to our whole construction is the theorems of Pringsheim and Landau. Here the crucial observation is that a series of positive terms (convergent or not) can be rearranged at will. Addition remains a *commutative* operation when the terms are positive. This is a sum of a *set* of quantities rather than the sum of a *sequence* of them.

The precise statement of the Pringsheim–Landau theorem is that, for a Dirichlet series with positive coefficients, the real boundary point of its convergence region must be a singularity.

Indeed this statement proves *itself* through the observation that $n^{a-z} = \sum_k \frac{(a-z)^k}{k!}(\log n)^k$ is a power series in $(a-z)$ with nonnegative coefficients. Thus the (unique) power series for $\sum \frac{a_n}{n^z} = \sum \frac{a_n}{n^a} n^{a-z}$ has positive coefficients in powers of $(\alpha - z)$. So let b be the real boundary point of the convergence region of $\sum \frac{a_n}{n^z}$, and suppose that b is a regular point and that $b < a$. Thus the power series in $(a-z)$ continues to converge a bit to the left of b and, by rearranging terms, the Dirichlet series converges there also, contradicting the meaning of b. A "natural" proof of a "natural" theorem follows, one with a very nice corollary which we record for future use.

(1) If a Dirichlet series with positive coefficients represents a function which is (can be continues to be) entire, then it is everywhere convergent.

Our ultimate aim is to prove that the L-series have no zeros on the line $\Re z = 1$. This is the nonvanishing of the L-series that we referred to in the chapter title. So let us begin with the simplest of all l-series, the ζ-function, $\zeta(z) = \sum \frac{1}{n^z}$. Our proof, in fact, was noticed by Narasimhan and is as follows: Assume, par contraire, that $\zeta(z)$ had a zero at $1 + ia$, a real. Then (sic!) the function $\zeta(z)\zeta(z + ia)$ would be entire. (See the appendix, page no. 651).

The only trouble points could be at $z = 1$ or at $z = 1 - ia$ where one of the factors has a pole, but these are then cancelled by the other factor, which, by our assumption, has a zero.

A bizarre conclusion, perhaps, that the Dirichlet series $\zeta(z)\zeta(z + ia)$ is entire. But how to get a contradiction? Surely there is no hint from its coefficients, they aren't even real. A*natural* step then would be to make them real by multiplying by the conjugate coefficient function, $\zeta(z)\zeta(z - ia)$, which of course is also entire. We are led, then, to form $\zeta^2(z)\zeta(z + ia)(z - ia)$.

This function is entire and has real coefficients, but are they positive? (We *want* them to be so that we can use (1).) Since these are complicated coefficients dependent on sums of complex powers of divisiors, we pass to the logarithm, $2 \log \zeta(z) + \log \zeta(z + ia) + \log \zeta(z - ia)$, which, by Euler's factorization of the ζ-function, has simple coefficients. A dangerous route, passing to the logarithm, because this surely destroys our everywhere analyticity. Nevertheless let us brazen forth (faint heart fair maiden never won).

By Euler's factorization, $2 \log \zeta(z) + \log \zeta(z + ia) + \log \zeta(z - ia) = \sum_p 2 \log \frac{1}{1-p^{-z}} + \log \frac{1}{1-p^{-z-ia}} + \log \frac{1}{1-p^{-z+ia}} = \sum_{p,v} \frac{1}{vp^{vz}} (2 + p^{-iva} + p^{+iva})$, and indeed these coefficients are nonnegative! The dangerous route is now reversed by exponentiating. We return to our entire function while preserving the nonnegativity of the coefficients. All in all, then,

(2) $\zeta^2(z)\zeta(z+ia)\zeta(z-ia)$ is an entire Dirichlet series with nonnegative coefficients. Combining this with (1) implies the unbelievable fact that

(3) the Dirichlet series for $\zeta^2(z)\zeta(z+ia)\zeta(z-ia)$ is everywhere convergent.

The falsity of (3) can be established in may ways, especially is we recall that the coefficients are all nonnegative. For example, the subseries corresponding to $n = \mathit{power}$ of 2 is exactly equal to $\frac{1}{(1-2^{-z})^2} \cdot \frac{1}{1-2^{-z-ia}} \cdot \frac{1}{1-2^{-z+ia}}$ which exceeds $\frac{1}{(1-2^{-z})^2} \cdot \frac{1}{4}$ along the positive axis and thereby guarantees divergence at $z = 0$. Q.E.D.

And so we have the promised natural proof of the nonvanishing of the ζ-function which can then lead to the natural proof of the prime number theorem. We must turn to the general L-series which holds the germ of the proof of the prime progression theorem. Dirichlet pointed out that the *natural* way to treat these progressions is not one progression at a time but all of the pertinent progressions of a given modulus simultaneously, for this leads to the underlying group and thence to its dual group, the group of characters. Let us look, for example, at the modulus 10. The pertinent progressions are $10k+1$, $10k+3$, $10k+7, 10k+9$, so that the group is the multiplicative group of 1,3,7,9 (mod 10). The characters are

$$\begin{aligned}
\chi_1 &: \chi_1(1) = 1, \chi_1(3) = 1, \chi_1(7) = 1, \chi_1(9) = 1,\\
\chi_3 &: \chi_3(1) = 1, \chi_3(3) = i, \chi_3(7) = -i, \chi_3(9) = -1,\\
\chi_7 &: \chi_7(1) = 1, \chi_7(3) = -i, \chi_7(7) = i, \chi_7(9) = -1,\\
\chi_9 &: \chi_9(1) = 1, \chi_9(3) = -1, \chi_9(7) = -1, \chi_9(9) = 1
\end{aligned},$$

and so the L-series are

$$L_1(z) = \prod_{p\equiv 1}\frac{1}{1-p^{-z}}\prod_{p\equiv 3}\frac{1}{1-p^{-z}}\prod_{p\equiv 7}\frac{1}{1-p^{-z}}\prod_{p\equiv 9}\frac{1}{1-p^{-z}},$$

$$L_3(z) = \prod_{p\equiv 1}\frac{1}{1-p^{-z}}\prod_{p\equiv 3}\frac{1}{1-ip^{-z}}\prod_{p\equiv 7}\frac{1}{1+ip^{-z}}\prod_{p\equiv 9}\frac{1}{1+p^{-z}},$$

$$L_7(z) = \prod_{p\equiv 1} \frac{1}{1-p^{-z}} \prod_{p\equiv 3} \frac{1}{1+ip^{-z}} \prod_{p\equiv 7} \frac{1}{1-ip^{-z}} \prod_{p\equiv 9} \frac{1}{1+p^{-z}},$$

and

$$L_9(z) = \prod_{p\equiv 1} \frac{1}{1-p^{-z}} \prod_{p\equiv 3} \frac{1}{1+p^{-z}} \prod_{p\equiv 7} \frac{1}{1+p^{-z}} \prod_{p\equiv 9} \frac{1}{1-p^{-z}}.$$

(Here $\Re z > 1$ to insure convergence and the subscripting of the characters is used to reflect the isomorphism of the dual group and the original group.)

The generating function for the primes in the arithmetic progressions ((mod 10) in this case) are then linear combinations of the logarithms of these L-series. And so indeed the crux is the nonvanishing of these L-series.

What could be more natural or more in the spirit of Dirichlet, but to prove these separate nonvanishings altogether? So we are led to take the product of all the L-series! (Landau uses the same device to prove nonvanishing of the L-series at point 1.)

The result is the Dirichlet series

$$Z(z) = \prod_{p\equiv 1} \frac{1}{(1-p^{-z})^4} \prod_{p\equiv 3} \frac{1}{(1-p^{-4z})} \times \prod_{p\equiv 7} \frac{1}{(1-p^{-4z})} \prod_{p\equiv 9} \frac{1}{(1-p^{-2z})^2},$$

and the problem reduces to showing that $Z(z)$ is zero-free on $\Re z = 1$.

Of course, this is equivalent to showing that $\prod_{p\equiv 1} \frac{1}{1-p^{-z}}$ is zero-free on $\Re z = 1$, which seems, at first glance, to be a more attractive form of the problem. This is misleading, however, and we are better off with $Z(z)$, which is the product of L-series and is an entire function except possibly for a simple pole at $z = 1$. (See the appendix.)

Guided by the special cases let us turn to the general one. So let A be a positive integer, and denote by G_A the multiplicative group of residue classes (mod A) which are prime to A. Set $h = \phi(A)$, and denote the group elements by $1 = n_1, n_2, \ldots, n_h$. Denote the dual group of G_A by $\hat{G}_A$ and its elements by $\chi_1, \chi_{n_2}, \ldots, \chi_{n_h}$ arranged so that $n_i \leftrightarrow \chi_{n_i}$ is an isomorphism of G and $\hat{G}$. Next, for $\Re z > 1$, write $L_{n_i}(z) = \prod_{n_j} \prod_{p \equiv n_j} \frac{1}{1-\chi_{n_i}(n_j)p^{-z}}$ and finally set $Z(z) = \prod_{n_i} L_{n_i}(z)$. As in the case $A = 10$, elementary algebra leads to $Z(z) = \prod_{n_j} \prod_{p \equiv n_j} \frac{1}{(1-p^{-h_j z})^{h/h_j}}$, where h_j is the order of the group element n_j.

As before, $Z(z)$ is entire except possibly for a simple pole at $z = 1$, and we seek a proof that $Z(i+ia) \neq 0$ for real a. So again we assume $Z(1+ia) = 0$, form $Z^2(z)Z(z+ia)Z(z-ia)$, and conclude that it is entire. We note that its logarithm and hence that it itself has nonnegative coefficients so that (1) is applicable.

So, with dazzling speed, we see that a zero of any L-series would lead to the everywhere convergence of the Dirichlet series (with nonnegative coefficients) $Z^2(z)Z(z+ia)Z(z-ia)$.

The end game (final contradiction) is also as before although 2 may not be among the primes in the resultant product, and we may have to take some other prime π. Nonetheless again we see that the subseries of powers of π diverges at $z = 0$ which gives us our QED.

Appendix. A proof that the L-series are everywhere analytic functions with the exception of the principal L-series, L_1 at the single point $z = 1$, which is a simple pole.

Lemma. *For any θ in [0,1), define $f(z) = \sum_{n=1}^{\infty} \frac{1}{(n-\theta)^z} - \frac{1}{z-1}$ for $\Re z > 1$. Then $f(z)$ is continuable to an entire function.*

PROOF. Since, for $\Re z > 1$, $\int_0^\infty e^{nt} e^{\theta t} t^{z-1} dt = \frac{1}{(n-\theta)^z} \int_0^\infty e^{-t} \times t^{z-1} dt = \frac{\Gamma(z)}{(n-\theta)^z}$, by summing, we get $\sum \frac{1}{(n-\theta)^z} = \frac{1}{\Gamma(z)} \int_0^\infty \frac{e^{\theta t}}{e^t - 1} \times t^{z-1} dt$ or $\sum \frac{1}{(n-\theta)^z} - \frac{1}{z-1} = \frac{1}{\Gamma(z)} \int_0^\infty \left(\frac{e^{\theta t}}{e^t-1} - \frac{e^{-t}}{t} \right) t^{z-1} dt$. Since $\frac{e^{\theta t}}{e^t-1} - \frac{e^{-t}}{t}$ is analytic and has integrable derivatives on $[0, \infty)$, we may integrate by parts repeatedly and thereby get $\sum \frac{1}{(n-\theta)^z} - \frac{1}{z-1} = \frac{1}{\Gamma(z+k)} \int_0^\infty (-\frac{d}{dt})^k \left(\frac{e^{\theta t}}{e^t-1} - \frac{e^{-t}}{t} \right) t^{z+k-1} dt$. This gives continuity to $\Re z > -k$, and, since k is arbitrary, the continuity is to the entire plane.

Problems for Chapter VI

1. Prove, by elementary methods, that there are infinitely many primes not ending in the digit 1.
2. Prove that there are infinitely many primes p for which neither $p+2$ nor $p-2$ is prime.
3. Prove that at least $1/6$ of the integers are *not* expressible as the sum of 3 squares.
4. Prove that $\Gamma(z)$ has no zeros in the whole plane, although, it has poles.
5. Suppose $\delta(x)$ decreases to 0 as $x \to \infty$. Produce an $\varepsilon(x)$ which goes to 0 at ∞ but for which $\delta(x\varepsilon(x)) = o(\varepsilon(x))$.

VII

Simple Analytic Proof of the Prime Number Theorem

The magnificent prime number theorem has received much attention and many proofs throughout the past century. If we ignore the (beautiful) elementary proofs of Erdös and Selberg and focus on the analytic ones, we find that they all have some drawbacks. The original proofs of Hadamard and de la Vallée Poussin were based, to be sure, on the nonvanishing of $\zeta(z)$ in $\Re z \geq 1$, but they also required annoying estimates of $\zeta(z)$ at ∞, because the formulas for the coefficients of the Dirichlet series involve integrals over *infinite* contours (unlike the situation for power series) and so effective evaluation requires estimates at ∞.

The more modern proofs, due to Wiener and Ikehara (and also Heins) get around the necessity of estimating at ∞ and are indeed based only on the appropriate nonvanishing of $\zeta(z)$, but they are tied to certain results of Fourier transforms. We propose to return to contour integral methods to avoid Fourier analysis and also to use finite contours to avoid estimates at ∞. Of course certain errors are introduced thereby, but the point is that these can be effectively minimized by elementary arguments.

So let us begin with the well-known fact about the well-known fact about the ζ-function (see Chapter 6, page 3)

$$(z-1)\zeta(z) \text{ is analytic and zero-free throughout } \Re z \geq 1. \qquad (1)$$

This will be assumed throughout and will allow us to give our proof of the prime number theorem.

In fact we give two proofs. This first one is the shorter and simpler of the two, but we pay a price in that we obtain one of Landau's equivalent forms of the theorem rather than the standard form, $\pi(N) \sim N/\log N$. Our second proof is a more direct assault on $\pi(N)$ but is somewhat more intricate than the first. Here we find some of Tchebychev's elementary ideas very useful.

Basically our novelty consists in using a modified contour integral,

$$\int_\Gamma f(z)N^z\left(\frac{1}{z}+\frac{z}{R^2}\right)dz,$$

rather than the classical one, $\int_C f(z)N^z z^{-1}dz$. The method is rather flexible, and we could use it to directly obtain $\pi(N)$ by choosing $f(z) = \log\zeta(z)$. We prefer, however, to derive both proofs from the following convergence theorem. Actually, this theorem dates back to Ingham, but his proof is á la Fourier analysis and is much more complicated than the contour integral method we now give.

Theorem. *Suppose* $|a_n| \le 1$, *and form the series* $\sum a_n n^{-z}$ *which clearly converges to an analytic function* $F(z)$ *for* $\Re z > 1$. *If, in fact,* $F(z)$ *is analytic throughout* $\Re z \ge 1$, *then* $\sum a_n n^{-z}$ *converges throughout* $\Re z \ge 1$.

PROOF OF THE CONVERGENCE THEOREM. Fix a w in $\Re w \ge 1$. Thus $F(z+w)$ is analytic in $\Re z \ge 0$. We choose an $R \ge 1$ and determine $\delta = \delta(R) > 0$, $\delta \le \frac{1}{2}$ and an $M = M(R)$ so that

$$F(z+w) \text{ is analytic and bounded by } M \text{ in } -\delta \le \Re z,\ |z| \le R. \tag{2}$$

Now form the counterclockwise contour Γ bounded by the arc $|z| = R$, $\Re z > -\delta$, and the segment $\Re z = -\delta$, $|z| \le R$. Also denote by A and B, respectively, the parts of Γ in the right and left half planes.

By the residue theorem,

$$2\pi i F(w) = \int_\Gamma F(z+w)N^z\left(\frac{1}{z}+\frac{z}{R^2}\right)dz.$$

Now on A, $F(z+w)$ is equal to its series, and we split this into its partial sum $S_N(z+w)$ and remainder $r_N(z+w)$. Again by the residue theorem,

$$\int_A S_N(z+w)N^z\left(\frac{1}{z}+\frac{z}{R^2}\right)dz$$
$$= 2\pi i S_N(w) - \int_{-A} S_N(z+w)N^z\left(\frac{1}{z}+\frac{z}{R^2}\right)dz,$$

with $-A$ as usual denoting the reflection of A through the origin. Thus, changing z to $-z$, this can be written as

$$\int_A S_N(z+w)N^z\left(\frac{1}{z}+\frac{z}{R^2}\right)dz$$
$$= 2\pi i S_N(w) - \int_{A} S_N(w-z)N^{-z}\left(\frac{1}{z}+\frac{z}{R^2}\right)dz \quad (4)$$

Combining (3) and (4) gives

$$2\pi[F(w) - S_N(w)]$$
$$= \int_A \left[r_N(z+w)N^z - \frac{S_N(w-z)}{N^z}\right]\left(\frac{1}{z}+\frac{z}{R^2}\right)dz \quad (5)$$
$$+ \int_B F(z+w)N^z\left(\frac{1}{z}+\frac{z}{R^2}\right)dz,$$

and, to estimate these integrals, we record the following (here as usual we write $\Re z = x$, and we use the notation $\alpha \ll \beta$ to mean simply that $|\alpha| \leq |\beta|$):

$$\frac{1}{z}+\frac{z}{R^2} = \frac{2x}{R^2} \text{ along } |z| = R \text{(in particular on)} A, \quad (6)$$

$$\frac{1}{z}+\frac{z}{R^2} \ll \frac{1}{\delta}\left(1+\frac{|z|^2}{R^2}\right) \leq \frac{2}{\delta} \text{ on the line } \Re z$$
$$= -\delta,\ |z| \leq R, \quad (7)$$

$$r_N(z+w) \ll \sum_{n=N+1}^{\infty} \frac{1}{n^{x+1}} \leq \int_N^{\infty} \frac{dn}{n^{x+1}} = \frac{1}{xN^x}, \quad (8)$$

and

$$S_N(w-z) \ll \sum_{n=1}^{N} n^{x-1} \leq N^{x-1} + \int_0^N n^{x-1}dn$$

$$= N^x \left(\frac{1}{N} + \frac{1}{x} \right). \tag{9}$$

By (6), (8), (9), on A,

$$\left[r_N(z+w)N^z - \frac{S_N(w-z)}{N^z} \right] \left(\frac{1}{z} + \frac{z}{R^2} \right)$$
$$\ll \left(\frac{1}{x} + \frac{1}{x} + \frac{1}{N} \right) \frac{2x}{R^2} \leq \frac{4}{R^2} + \frac{2}{RN},$$

and so, by the "maximum times length" estimate (M–L formula) for integrals, we obtain

$$\int_A \left[r_N(z+w)N^z - \frac{S_N(w-z)}{N^z} \right] \left(\frac{1}{z} + \frac{z}{R^2} \right) dz \ll \frac{4\pi}{R} + \frac{2\pi}{N}. \tag{10}$$

Next, by (2), (6), and (7), we obtain

$$\int_B F(z+w)N^z \left(\frac{1}{z} + \frac{z}{R^2} \right) dz$$
$$\ll \int_{-R}^{R} M \cdot N^{-\delta} \frac{2}{\delta} dy + 2M \int_{-\delta}^{0} n^x \frac{2|x|}{R^2} \frac{3}{2} dx \tag{11}$$
$$\leq \frac{4MR}{\delta N^\delta} + \frac{6M}{R^2 \log^2 N}.$$

Inserting the estimates (10) and (11) into (5) gives

$$F(w) - S_N(w) \ll \frac{2}{R} + \frac{1}{N} + \frac{MR}{\delta N^\delta} + \frac{M}{R^2 \log^2 N},$$

and, if we fix $R = 3/\epsilon$, we note that this right-hand side is $< \epsilon$ for all large N. We have verified the very definition of convergence!

First Proof of the Prime Number Theorem.

Following Landau, we will show that the convergence of $\sum_n \frac{\mu(n)}{n}$ (as given above) implies the PNT. Indeed all we need about this convergent series is the simple corollary that $\sum_{n \leq N} \mu(n) = o(N)$.

Expressing everything in terms of the ζ-function, then, we have established the fact that $\frac{1}{\zeta(z)}$ has coefficients which go to 0 on average.

The PNT is equivalent to the fact that the average of the coefficients of $\frac{\zeta'}{\zeta}(z)$ is equal to 1. For simply note that

$$\begin{aligned}\frac{\zeta'}{\zeta}(z) &= \frac{d}{dz}\log\zeta(z) = \frac{d}{dz}\log\prod_p \frac{1}{1-p^{-z}}\\ &= \frac{d}{dz}\sum_p \log\frac{1}{1-p^{-z}} = \sum_p \frac{d}{dz}\log\frac{1}{1-p^{-z}}\\ &= \sum_p \frac{\log p\cdot p^{-z}}{1-p^{-z}} = \sum_p \frac{\log p}{p^z-1}.\end{aligned}$$

This last series is the same as $\sum \frac{A(n)}{n^z}$ where $A(n)$ is $\log p$ whenever n is a power of p, p any prime, and 0 otherwise. So indeed the average of these coefficients is $\frac{1}{N}\sum_{n\le N} A(n)$ whose limit being 1 is exactly the prime number theorem.

In short, we want the average value of the coefficients of $\frac{\zeta'}{\zeta}(z) - \zeta(z)$ to approach 0. Writing this function as

$$\frac{1}{\zeta(z)}[\zeta'(z) - \zeta^2(z)] = \sum \frac{\mu(n)}{n^2}\left[\sum \frac{\log n}{n^z} - \sum \frac{d(n)}{n^z}\right],$$

we may write this average (of the first N terms)as

$$\begin{aligned}&\frac{1}{N}\sum_{ab\le N}\mu(a)[\log b - d(b)]\\ &\quad = \frac{1}{N}\sum_{ab\le N}\mu(a)[\log b - d(b) + 2\gamma] - \frac{2\gamma}{N},\end{aligned}$$

where 2γ is chosen as the constant for which

$$\sum_{b=1}^{K}[\log b - d(b) + 2\gamma]$$

becomes $O(\sqrt{K})$.

Now we use the Landau corollary that $\sum \mu(n) = o(N)$ to conclude that

$$\frac{1}{N}\sum \mu(n) \ll \delta(N),$$

where $\delta(N)$ *decreases* to 0, and our trick is to pick a function $w(N)$ which approaches ∞ but such that $w(N)\delta\left[\frac{N}{w(N)}\right]$ approaches 0.

This done, we may conclude that

$$\sum_{n\leq N} A(n) = N + O\left[\frac{N}{\sqrt{w(N)}}\right] + O\left[Nw(N)\delta\left[\frac{N}{w(N)}\right]\right]$$

$$= N + O(N),$$

and the proof is complete.

Second Proof of the Prime Number Theorem.

In this section, we begin with Tchebychev's observation that

$$\sum_{p\leq n} \frac{\log p}{p} - \log n \quad \text{is bounded,} \tag{12}$$

which he derives in a direct elementary way from the prime factorization on $n!$

The point is that the prime number theorem is easily derived from

$$\sum_{p\leq n} \frac{\log p}{p} - \log n \quad \text{converges to a limit,} \tag{13}$$

by simple summation by parts, which we leave to the reader. Nevertheless the transition from (12) to (13) is not a simple one, and we turn to this now.

So, for $\Re z > 1$, form the function

$$f(z) = \sum_{n=1}^{\infty} \frac{1}{n^z}\left(\sum_{p\leq n} \frac{\log p}{p}\right) = \sum_{p} \frac{\log p}{p}\left[\sum_{n\geq p} \frac{1}{n^z}\right].$$

Now

$$\sum_{n\geq p} \frac{1}{n^z} = \frac{1}{(z-1)p^{z-1}} + z\int_p^{\infty} \frac{1-\{t\}}{t^{z+1}}\,dt$$

$$= \frac{p}{(z-1)}\left(\frac{1}{p^z-1} + A_p(z)\right)$$

where $A_p(z)$ is analytic for $\Re z > 0$ and is bounded by

$$\frac{1}{p^x(p^x - 1)} + \frac{|z(z-1)|}{xp^{x+1}}.$$

Hence,

$$f(z) = \frac{1}{z-1}\left[\sum_p \frac{\log p}{p^z - 1} + A(z)\right],$$

where $A(z)$ is analytic for $\Re z > \frac{1}{2}$ by the Weierstrass M-test.

By Euler's factorization formula, however, we recognize that

$$\sum_p \frac{\log p}{p^z - 1} = \frac{-d}{dz} \log \zeta(z), \tag{14}$$

and so we deduce, by (1), that $f(z)$ is analytic in $\Re z \le 1$ except for a double pole with principal part $1/(z-1)^2 + c/(z-1)$ at $z = 1$. Thus if we set

$$F(z) = f(z) + \zeta'(z) - c\zeta(z) = \sum \frac{a_n}{n^z}$$

where

$$a_n = \sum_{p \le n} (\log p)/p - \log n - c,$$

we deduce that $F(z)$ is analytic in $\Re z \ge 1$.

From (12) and our convergence theorem, then, we conclude that

$$\sum \frac{a_n}{n} \quad \text{converges,}$$

and from this and the fact, from (14), that $a_n + \log n$ is nondecreasing, we proceed to prove $a_n \to 0$.

By applying the Cauchy criterion we find that, for N large,

$$\sum_N^{N(1+\epsilon)} \frac{a_n}{n} \le \epsilon^2 \tag{15}$$

and

$$\sum_{N(1-\epsilon)}^{N} \frac{a_n}{n} \ge -\epsilon^2. \tag{16}$$

In the range N to $N(1+\epsilon)$, by (14), $a_n \geq a_N + \log(N/n) \geq a_N - \epsilon$. So $\sum_N^{N(1+\epsilon)} a_n/n \geq (a_N - \epsilon) \sum_N^{N(1+\epsilon)} 1/n$, and (15) yields

$$a_n \ll \epsilon + \frac{\epsilon^2}{\sum_N^{N(1+\epsilon)} \frac{1}{n}} \ll \epsilon + \frac{\epsilon^2}{N\epsilon/N(1+\epsilon)} = 2\epsilon + \epsilon^2. \quad (17)$$

Similarly in $[N(1-\epsilon), N]$, $a_n \leq a_N + \log(N/n) \leq a_N + \epsilon/(1-\epsilon)$, so that

$$\sum_{N(1-\epsilon)}^{N} \frac{a_n}{n} \leq \left(a_N + \frac{\epsilon}{1-\epsilon}\right) \sum_{N(1-\epsilon)}^{N} \frac{1}{n},$$

and (16) gives

$$a_N \geq \frac{-\epsilon}{1-\epsilon} - \frac{\epsilon^2}{\sum_{N(1-\epsilon)}^{N} \frac{1}{n}} \geq \frac{-\epsilon}{1-\epsilon} - \frac{\epsilon^2}{N\epsilon/N} = \frac{\epsilon^2 - 2\epsilon}{1-\epsilon}. \quad (18)$$

Taken together, (17) and (18) establish that $a_N \to 0$, and so (13) is proved.

Problems for Chapter VII

1. Given that $\sum \frac{a_n}{n}$ converges, prove that $\sum_{n=1}^{N} a_n = o(N)$.

2. Given that $\sum \frac{a_n}{n}$ converges and that $a_n - a_{n-1} > \frac{-1}{n}$, prove that $a_n \to 0$.

3. Show that $d(n)$, the number of divisors of n, is $O(n^{\varepsilon})$ for every positive ε.

4. In fact, show that $d(n) \ll n^{\frac{1}{\log \log n}}$.

Index

Addition problems, 1–2
Affine property, 41
Analytic functions, L-series as, 63
Analytic method, 1
Analytic number theory, 1–14
Analytic proof of prime number theorem, 65–71
Approximation lemma, basic, 42–47
Arithmetic progressions, 41
  dissection into, 14
  sequences without, 41–47
Asymptotic formula, 4

Basic approximation lemma, 42–47

Cauchy criterion, 71
Cauchy integral, 23–24
Cauchy's theorem, 18–19
Change making, 2–5
Commutative operation, 59
Complex numbers, 18
Contour integral, modified, 66
Contour integration, 46
Contours
  finite, 65
  infinite, 65
Convergence theorem, 66
  proof of, 66–68
Crazy dice, 5–8

Dice, crazy, 5–8
Dirichlet series, 59–60, 62
Dirichlet theorem, 45, 50
Dissection into arithmetic progressions, 14

Elliptic integral, 33
Entire functions, 60
Erdös, Paul, vii
Erdös-Fuchs theorem, 31, 35–38
Euler's factorization, 60
Euler's factorization formula, 71
Euler's theorem, 11–12
Evens and odds, dissection into, 14
Extremal sets, 42

Finite contours, 65
Fourier analysis, 65

Generation functions, 1
  of asymptotic formulas, 18–19
  of representation functions, 7

Infinite contours, 65
Integers, 1
  breaking up, 17
  nonnegative, splitting, 8–10

L-series
  as analytic functions, 63
  general, 61–62

nonvanishing of, see Nonvanishing of L-series
zero of any, 63
Lagrange theorem, 49
Landau corollary, 69
L'Hôpital's rule, 5

"Magnitude property," 53
Mathematics, vii
"Monotone majorant," 45

"Natural" proof, 59
of nonvanishing of L-series, 59–63
Nonnegative integers, splitting, 8–10
Nonvanishing of L-series, 60
"natural" proof of, 59–63

Odds and evens, dissection into, 14

Parseval upper bound, 36
Parseval's identity, 33–34
Partial fractional decomposition, 3–4
Partition function, 17–29
Permission constant, 42
Pigeonhole principle, 50
PNT, see Prime number theorem
Prime number theorem (PNT), 65
analytic proof of, 65–71
first proof of, 68–70
second proof of, 70–72
Pringsheim-Landau theorem, 59
Progressions, arithmetic, see Arithmetic progressions

$q(n)$, coefficients of, 25–29

Relative error, 4
Representation functions, 7
generation functions of, 7
near constancy of, 31
Riemann integral, 20
double, 31
Riemann sums, 20–25
Roth theorem, 46–47
Rulers, marks on, 12–13

Schnirelmann's theorem, 50–51
Schwarz inequality, 34
Sequences without arithmetic progressions, 41–47
Splitting problem, 8–10
Stirling's formula, 4, 27, 29
Szemeredi-Furstenberg result, 43

Taylor coefficients, 3
Tchebychev's observation, 70

Unit circle, 13

Waring problem, 49–56
Weyl sums, 51–52

Graduate Texts in Mathematics

continued from page ii

61 WHITEHEAD. Elements of Homotopy Theory.
62 KARGAPOLOV/MERLZJAKOV. Fundamentals of the Theory of Groups.
63 BOLLOBAS. Graph Theory.
64 EDWARDS. Fourier Series. Vol. I 2nd ed.
65 WELLS. Differential Analysis on Complex Manifolds. 2nd ed.
66 WATERHOUSE. Introduction to Affine Group Schemes.
67 SERRE. Local Fields.
68 WEIDMANN. Linear Operators in Hilbert Spaces.
69 LANG. Cyclotomic Fields II.
70 MASSEY. Singular Homology Theory.
71 FARKAS/KRA. Riemann Surfaces. 2nd ed.
72 STILLWELL. Classical Topology and Combinatorial Group Theory. 2nd ed.
73 HUNGERFORD. Algebra.
74 DAVENPORT. Multiplicative Number Theory. 2nd ed.
75 HOCHSCHILD. Basic Theory of Algebraic Groups and Lie Algebras.
76 IITAKA. Algebraic Geometry.
77 HECKE. Lectures on the Theory of Algebraic Numbers.
78 BURRIS/SANKAPPANAVAR. A Course in Universal Algebra.
79 WALTERS. An Introduction to Ergodic Theory.
80 ROBINSON. A Course in the Theory of Groups. 2nd ed.
81 FORSTER. Lectures on Riemann Surfaces.
82 BOTT/TU. Differential Forms in Algebraic Topology.
83 WASHINGTON. Introduction to Cyclotomic Fields. 2nd ed.
84 IRELAND/ROSEN. A Classical Introduction to Modern Number Theory. 2nd ed.
85 EDWARDS. Fourier Series. Vol. II. 2nd ed.
86 VAN LINT. Introduction to Coding Theory. 2nd ed.
87 BROWN. Cohomology of Groups.
88 PIERCE. Associative Algebras.
89 LANG. Introduction to Algebraic and Abelian Functions. 2nd ed.
90 BRØNDSTED. An Introduction to Convex Polytopes.
91 BEARDON. On the Geometry of Discrete Groups.
92 DIESTEL. Sequences and Series in Banach Spaces.
93 DUBROVIN/FOMENKO/NOVIKOV. Modern Geometry—Methods and Applications. Part I. 2nd ed.
94 WARNER. Foundations of Differentiable Manifolds and Lie Groups.
95 SHIRYAEV. Probability. 2nd ed.
96 CONWAY. A Course in Functional Analysis. 2nd ed.
97 KOBLITZ. Introduction to Elliptic Curves and Modular Forms. 2nd ed.
98 BRÖCKER/TOM DIECK. Representations of Compact Lie Groups.
99 GROVE/BENSON. Finite Reflection Groups. 2nd ed.
100 BERG/CHRISTENSEN/RESSEL. Harmonic Analysis on Semigroups: Theory of Positive Definite and Related Functions.
101 EDWARDS. Galois Theory.
102 VARADARAJAN. Lie Groups, Lie Algebras and Their Representations.
103 LANG. Complex Analysis. 3rd ed.
104 DUBROVIN/FOMENKO/NOVIKOV. Modern Geometry—Methods and Applications. Part II.
105 LANG. $SL_2(\mathbf{R})$.
106 SILVERMAN. The Arithmetic of Elliptic Curves.
107 OLVER. Applications of Lie Groups to Differential Equations. 2nd ed.
108 RANGE. Holomorphic Functions and Integral Representations in Several Complex Variables.
109 LEHTO. Univalent Functions and Teichmüller Spaces.
110 LANG. Algebraic Number Theory.
111 HUSEMÖLLER. Elliptic Curves.
112 LANG. Elliptic Functions.
113 KARATZAS/SHREVE. Brownian Motion and Stochastic Calculus. 2nd ed.
114 KOBLITZ. A Course in Number Theory and Cryptography. 2nd ed.
115 BERGER/GOSTIAUX. Differential Geometry: Manifolds, Curves, and Surfaces.
116 KELLEY/SRINIVASAN. Measure and Integral. Vol. I.
117 SERRE. Algebraic Groups and Class Fields.
118 PEDERSEN. Analysis Now.

119 ROTMAN. An Introduction to Algebraic Topology.
120 ZIEMER. Weakly Differentiable Functions: Sobolev Spaces and Functions of Bounded Variation.
121 LANG. Cyclotomic Fields I and II. Combined 2nd ed.
122 REMMERT. Theory of Complex Functions. *Readings in Mathematics*
123 EBBINGHAUS/HERMES et al. Numbers. *Readings in Mathematics*
124 DUBROVIN/FOMENKO/NOVIKOV. Modern Geometry—Methods and Applications. Part III.
125 BERENSTEIN/GAY. Complex Variables: An Introduction.
126 BOREL. Linear Algebraic Groups. 2nd ed.
127 MASSEY. A Basic Course in Algebraic Topology.
128 RAUCH. Partial Differential Equations.
129 FULTON/HARRIS. Representation Theory: A First Course. *Readings in Mathematics*
130 DODSON/POSTON. Tensor Geometry.
131 LAM. A First Course in Noncommutative Rings.
132 BEARDON. Iteration of Rational Functions.
133 HARRIS. Algebraic Geometry: A First Course.
134 ROMAN. Coding and Information Theory.
135 ROMAN. Advanced Linear Algebra.
136 ADKINS/WEINTRAUB. Algebra: An Approach via Module Theory.
137 AXLER/BOURDON/RAMEY. Harmonic Function Theory.
138 COHEN. A Course in Computational Algebraic Number Theory.
139 BREDON. Topology and Geometry.
140 AUBIN. Optima and Equilibria. An Introduction to Nonlinear Analysis.
141 BECKER/WEISPFENNING/KREDEL. Gröbner Bases. A Computational Approach to Commutative Algebra.
142 LANG. Real and Functional Analysis. 3rd ed.
143 DOOB. Measure Theory.
144 DENNIS/FARB. Noncommutative Algebra.
145 VICK. Homology Theory. An Introduction to Algebraic Topology. 2nd ed.
146 BRIDGES. Computability: A Mathematical Sketchbook.
147 ROSENBERG. Algebraic K-Theory and Its Applications.
148 ROTMAN. An Introduction to the Theory of Groups. 4th ed.
149 RATCLIFFE. Foundations of Hyperbolic Manifolds.
150 EISENBUD. Commutative Algebra with a View Toward Algebraic Geometry.
151 SILVERMAN. Advanced Topics in the Arithmetic of Elliptic Curves.
152 ZIEGLER. Lectures on Polytopes.
153 FULTON. Algebraic Topology: A First Course.
154 BROWN/PEARCY. An Introduction to Analysis.
155 KASSEL. Quantum Groups.
156 KECHRIS. Classical Descriptive Set Theory.
157 MALLIAVIN. Integration and Probability.
158 ROMAN. Field Theory.
159 CONWAY. Functions of One Complex Variable II.
160 LANG. Differential and Riemannian Manifolds.
161 BORWEIN/ERDÉLYI. Polynomials and Polynomial Inequalities.
162 ALPERIN/BELL. Groups and Representations.
163 DIXON/MORTIMER. Permutation Groups.
164 NATHANSON. Additive Number Theory: The Classical Bases.
165 NATHANSON. Additive Number Theory: Inverse Problems and the Geometry of Sumsets.
166 SHARPE. Differential Geometry: Cartan's Generalization of Klein's Erlangen Program.
167 MORANDI. Field and Galois Theory.
168 EWALD. Combinatorial Convexity and Algebraic Geometry.
169 BHATIA. Matrix Analysis.
170 BREDON. Sheaf Theory. 2nd ed.
171 PETERSEN. Riemannian Geometry.
172 REMMERT. Classical Topics in Complex Function Theory.
173 DIESTEL. Graph Theory.
174 BRIDGES. Foundations of Real and Abstract Analysis.
175 LICKORISH. An Introduction to Knot Theory.
176 LEE. Riemannian Manifolds.
177 NEWMAN. Analytic Number Theory.
178 CLARKE/LEDYAEV/STERN/WOLENSKI. Nonsmooth Analysis and Control Theory.